Mohamed Atta

Corn Breeding for Drought Tolerance

AF536927

Mohamed Atta

Corn Breeding for Drought Tolerance

Noor Publishing

Imprint

Any brand names and product names mentioned in this book are subject to trademark, brand or patent protection and are trademarks or registered trademarks of their respective holders. The use of brand names, product names, common names, trade names, product descriptions etc. even without a particular marking in this work is in no way to be construed to mean that such names may be regarded as unrestricted in respect of trademark and brand protection legislation and could thus be used by anyone.

Cover image: www.ingimage.com

Publisher:
Noor Publishing
is a trademark of
International Book Market Service Ltd., member of OmniScriptum Publishing Group
17 Meldrum Street, Beau Bassin 71504, Mauritius

Printed at: see last page
ISBN: 978-620-2-35308-3

Copyright © Mohamed Atta
Copyright © 2018 International Book Market Service Ltd., member of OmniScriptum Publishing Group
All rights reserved. Beau Bassin 2018

CORN BREEDING FOR DROUGHT TOLERANCE

Mohamed Mohamed Mohamed Atta
Agronomy Department, Faculty of Agriculture, Cairo University, Egypt.

INTRODUCTION

Corn or maize (*Zea mays* L.) is the third important cereal crop after wheat and rice. Generally, corn is used for human consumption, livestock and poultry feed, as well as, manufacturing starch and cooking oils. Corn is also grown for fodder and silage. The horizontal expansion of maize production in Egypt is only possible by growing maize in the new-reclaimed lands, which are mostly sandy soils with limiting water assets. This would expose the corn plants to water stress, which could result in obtaining low grain yields under such conditions. Moreover, the expected future shortage in irrigation water requires corn breeders to pay a great attention to develop tolerant maize cultivars to proper with these conditions. These cultivars could give high grain yield under both water deficit and well-water conditions. In maize, drought reduces leaf area, leaf chlorophyll contents, photosynthesis and ultimately reduces the grain yield (Athar and Ashraf, 2005). Also, drought at flowering widens the anthesis-silking interval (ASI) in maize, which hardly reduces the kernel set (Edmeadeas *et al.*, 2000). Drought is also accelerating leaf senescence that reduces the canopy size and severely affecting the crop yield (Moony and Duplesis, 1970). However, delayed leaf senescence could be affected positively for reducing the harmful effects of drought on crop yield (Rivero *et al.*, 2007).

Several authors (Westgate and Grant 1989, Chapman *et al*. 1996, Ribaut *et al*. 1997, Al-Naggar *et al*. 2000, Salih *et al*. 2014 , Atta and Masri 2015 and Atta *et al*. 2017) reported that maize is susceptible to water deficit during flowering stages, but it becomes less sensitive as reproduction progress (Classen and Shaw 1970, Westgate and Boyer 1985, Westgate and Grant 1989 , Al-Naggar *et al*. 2000 and Atta *et al*. 2017). Denmead and Shaw (1960) noted that water stress during the vegetative, silking and grain filling stages of corn production reduced grain yield by 25%, 50% and 21%, respectively. Classen and Shaw (1970) and Shaw (1977) reported that sensitive period of corn to drought stress extended from around one week before to two weeks after 50% silking. Shaw (1977) added that yield losses per day of comparable stress, before and after flowering, were around 45 and 60%, respectively, with the maximum yield losses at silking itself. Although yields most severely reduced by 70% due to stress coinciding with silking, yields were reduced by 40-50% from stresses occurring in the period 10-31 days after mid-silk as well as reducing in kernel number due to stresses occurring up to 22 days after silking (Grant *et al*. 1989). In addition, Nesmith and Ritchie (1992) observed that kernel numbers per plant were reduced by 8-20% and kernel weight declined by 21-25% when the plants were stressed in the period 18-31 days after silking. Atta *et al*. (2017) observed that water deficit caused a significant yield reduction of 54.60% at flowering and 47.98% at grain filling; such reduction was accompanied with significant reduction in all yield components except for rows/ear. They added that the highest reduction was observed by kernels/ear at flowering and 100-kernel weight and shelling % at grain filling.

Genotypic differences in anthesis-silking interval (ASI), number of ears per plant and kernels/ear would help plant breeder in initiating a successful breeding program to improve such a complicated character. Tolerant genotypes of corn were characterized by having shorter ASI, higher number of ears/plant and higher number of kernels/ear than susceptible ones (Hall *et al*. 1982 , Edmeades *et al*. 1993, Bolanos and Edmeades 1996, Ribaut *et al*. 1997 and Al-Naggar *et al*. 2004, 2008 a, 2011 and 2016). Numerous studies highlighted the role of corn genotypes in drought tolerance. There are prospect confirmation that hybrids possess their advantage over open-pollinated varieties and inbred lines in both stress and non-stress conditions as reported by Vasal *et al*. (1997), El-Ganayni *et al*. (2000) and Al-Naggar *et al*. (2008 a, 2011 and 2016). Also, O'Neill *et al*. (2004) showed considerable genetic distinction in the response of commercial hybrids to water stress executed through reproductive growth.

In order to start a successful breeding program for improving drought tolerance, available germplasm should be tested under water deficit conditions to indentify the best ones which could be used as source material for extracting drought tolerant inbred lines for developing drought tolerant single and three-way cross hybrids. Type of gene action, heritability and expected genetic advance from selection are prerequisites for starting a breeding program for developing a drought tolerant variety of maize (Al-Naggar and Atta, 2017). Literature review reveals that little research has been directly focused on studying the mode of gene action controlling yield under drought. Some researchers found that additive genetic effects play a major role in conditioning grain yield under water stress in tropical (Chapman and Edmeades, 1999 and Dhliwayo *et al*. 2009) and temperate (Betran *et al*. 2003) maize germplasm. Al-Naggar and Atta (2017) observed that the magnitude of additive was much higher than dominance variance for all studied traits, except for rows/ear under well watering and drought stress conditions. Response to selection for yield in populations under water stress conditions has also been reported (Al-Naggar *et al.* 2008 a and b), suggesting that additive gene action might be important in controlling yield. In addition, Derera *et al.* (2008) found also that non-additive gene action playing important roles in controlling grain yield under both water stress and favorable growing environments. Several authors (Bolanos and Edmeades 1996, Betran *et al.* 2003, Campos *et al.* 2004, Xiong *et al.* 2006 and Atta *et al.* 2017) reported significance of anthesis-silking interval, silk emergence, anthesis date and number of ears plant^{-1} in breeding drought tolerance in maize. Many investigators reported a decline in heritability for grain yield under stress (Rosielle and Hamblin 1981 and Blum 1988). Furthermore, it should be taken into consideration that the estimate of heritability applies only to environments sampled (Dudley and Moll 1969).

Although a wide array of biometrical tools is available to breeders, diallel analysis is one of the best biometrical tools to achieve the appropriate breeding methodology by characterizing genetic control of economically important traits. Upon diallel analysis, if the additive variance was more pronounced than non-additive effect in the inheritance of studied traits, then selection breeding would be very efficient for improving such traits. On the contrary, if non-additive variance was more pronounced than additive effect in the inheritance of studied traits, then heterosis breeding or hybrid breeding would be very efficient in this case.

The present work will discuss some of the problems may be faced the corn breeder concerning the choice of appropriate experimental design, data handling and methods of assessment of drought tolerance as well as the choice of appropriate breeding method for the development of drought-tolerant genotypes.

Controlling of Maize Pollination

- **Inflorescence**

- Maize has a separate male and female inflorescence, the plant is monocious and the flower is unisexual. It is a cross-pollinated crop and a diploid plant (2n=20).

- The male inflorescence is at the top of the plant and is called tassel which it carries staminate flowers.

- The female inflorescence is usually located in the middle of the stalk and is called ear which it carries the pistillate flowers. The ear originates from the axillary bud apice. The plant usually carries more than one ear (prolific).

- The male florets usually mature before the female florets (protandry). The difference in maturity of male and female florets could be affected by environmental conditions such as drought and heat stress.

- Pollen shed begins in the main tassel branch from the central spike or rachis after anther exertion.

- Spikelets are carried on the tassel in pairs, one of them is pedicellate and the other is sessile. Each spikelet has a pair of glumes and two florets. Each floret enclosed with lemma and palea. Each floret has three anthers, two of them are located adjacent to the palea and the third one is located adjacent to the lemma.

- The number of pollen grains produced by the tassel can be reached 25 million pollen grains on the average.

- Silk emerges from the top of ear husk and is functional stigmas. There is one stigma for each kernel. Silk emergence starts from the bottom to the tip of the ear. Drought may cause silk growth to stop and widens the anthesis-silking interval (ASI), which severely reduces the kernel set.

- The ear has multiple rows of paired spikelets. Each spikelet has two florets (the upper floret and the lower floret) but only the upper floret is developed. The upper floret has an ovary with an elongated style (silk) covered with feathery hairs.
- Fertilization is accomplished when pollen grains fertilize the silk and unite with the female gametes. Good synchronization of the time of pollen and silk emergence is required to obtain a good kernel set.

- **Artificial Hybridization and Self Pollination**

It should be used tassel and shoot bags to cover male and female inflorescence in order to avoid fertilization by unknown pollen grains. The method used in making self as well as cross pollination should be done as the following steps:

1- Covering the shoot by using glassine bags before the silks emergence to prevent the female gametes from fertilization by unknown pollen (Fig. 1).

2- Determine tassels for viable pollen present.

3- When the tassels and ears are at adequate maturity, they are prepared for pollination.

4- Covering the tassel with a kraft paper for the pollen source and cutting ear tips on the day before pollination (Fig. 2).

5- Pollination is carried out after one day of male and female prepared. Selfing is carried out by using the tassels on the same plants of covered ear shoot. On the contrary, cross-pollination is carried out by using the tassels from different plants.

6- The tassel bags should be secured on the ear shoot and fixed to the stalk until harvest.

- **Test of Viability of Maize Pollen Grains**

Success of the hybridization is required for the genetic improvement in breeding programs. Borém and Miranda (2007) reported that the viability of pollen grains is essential precondition for obtaining hybrid vigor genotypes and a good fixation of the fruit. In addition, it has a great importance for genetic improvement programs in various types of controlled pollination. Ferreira *et al.* (2007) reported that in maize the release of pollen grains can start from sunrise until noon, which it depending on the temperature, humidity and genetic constitution of the plant. They added that pollen grains in maize can loss viability within a range of one to four hours after being released into the air. Furthermore, Kaefer *et al.* (2016) observed that the best results of viable pollen grains of maize were obtained at 09:00 a.m.

There are various techniques used to assess the viability of pollen grains, involving germination *in vivo* and *in vitro* as well as the chemical dyes test which is based on cytological criteria such as coloration (Almeida *et al.* 2011). The germination *in vitro* technique in culture medium emulates the style-stigma conditions that inducing germination of the pollen tube (Kaefer *et al.* 2016). In order to obtain a good germination of the pollen grains, a specific formulation of the culture medium is required. Sucrose is consider the essential element in the in the culture medium, while the boron as boric acid and calcium as calcium dehydrate can maximize the medium efficiency. In addition, to give consistency to the medium and avoid damage to the pollen tube during evaluation, the agar must be used (Ferreira *et al.* 2007). Regarding to the chemical dyes test, the stain ability is considered simple procedure, inexpensive and provides quickly results. There are various dyes may be employed for the test of viability such as acetic carmine, triphenyltetrazolium chloride, aniline blue and malachite green with acid fuchsin (Kaefer *et al.* 2016). The estimation of pollen viability is given by counting the aborted and not aborted pollen showing stained and unstained, respectively (Alvim 2008). Regarding to the *in vitro* technique, Almeida *et al.* (2002) observed that *in vitro* technique of pollen grains influenced by environmental conditions such as temperature and relative humidity during the collection and the maturity phase of the tassel, which the newly formed gametes are more viable than pollen grains matured. Also, Kaefer *et al.* (2016) observed that the ambient temperature and relative humidity were the main influencing factors on pollen viability. The culture medium of *in vitro* germination of maize pollen grains composed of Sucrose 15%, Boric acid 0.01%, Calcium Nitrate 0.025% and Agar 0.6% with pH 6.0 according to Kaefer *et al.* (2016). They used the previously described culture medium to determine the germination percentage of maize pollen grains; the incubation temperature was 25 °C in a Biochemical Oxygen Demand (BOD). They counted the number of germinated grains after two hours of incubation. The germinated pollen grains count was done with the aid of optical microscope with a 10x objective increase by evaluating four fields of view, corresponding to four replications. In each field of view there were on average 40 pollen grains. They were considered germinated grains, which had pollen tubes which exceeded the length of the diameter of pollen grain itself.

Fig. 1. (A) Tassel emergence. (B) Tassel branch of corn showing anthers exertion. (C) Ear shoots of corn emerging from the leaf sheath [this stage is the convenient stage for shoot bagging). (D) Ear shoot after removing its husks, silks are attached to the tip of each ovary.

Fig. 2. Steps of selfing and crossing (A) ear shoot bagging with glassine bag to prevent pollination. (B) Ear shoot cut back on the day previous to pollination. (C) Tassel bag is placed over the tassel.

Fig. 2 Cont. (D) Silk brush is dusted by pollen grains. (F) The shoot is protected after pollination by covering with tassel bag. All images were taken at the Agricultural Research and Experiment Station, of Faculty of Agriculture, Cairo University, Giza, by Dr. Mohamed Atta, Dr. Ramadan Elbadawy and Dr. Mahmoud Mabrouk.

Methods of Corn Breeding

In order to determine the appropriate breeding method for a set of genotypes, it should be characterized genetic control of economically important traits under drought. Although a wide array of biometrical tools is available to breeders, diallel analysis is one of the best biometrical tools to achieve that. Upon diallel analysis, if the additive variance was more pronounced than non-additive effect in the inheritance of studied traits, then selection breeding would be very efficient for improving such traits. On the contrary, if non-additive variance was more pronounced than additive effect in the inheritance of studied traits, then heterosis breeding or hybrid corn breeding would be very efficient in this case.

- ## Source of Plant Materials

It can be distinguished different types of plant materials depending on the degree of selection that plant materials undergone, as illustrated below:

(1) Land races:

Land races are local varieties developed by farmers during a lot of generations of selection without applying usual procedures of plant breeding. It was the first material used in the early generation of plant breeding. It could still be useful for breeding programs in many areas. Reid Yellow Dent and Lancaster Sure Crop are two examples of the well known local varieties of maize.

(2) Introduced materials:

Genetic materials can be included introduced materials from seed introduction centers such as CIMMYT (International Maize and Wheat Improvement Center).

(3) Composite or synthetics:

Composite or synthetics are developed by breeders by recombining cultivars, inbred lines or varieties *e.g.* Stiff Stalk Synthetic in maize.

(4) Recurrent selection populations:

Recurrent selection populations are selected populations by evaluating genotypes and recombining the selected ones. They should be adapted to a wide range of environments and have enough frequency of favorable genes.

(5) Elite lines:

Elite lines are developed from selected breeding material. They are used for transferring favorable alleles to other genotype or they are used as adapted stocks to develop new improved lines by backcrossing.

(6) Selected crosses:

Selected crosses are developed from elite lines. They are used to develop new elite lines by pedigree method.

Selection Methods

- Mass Selection

Mass selection was one of the earliest breeding procedures practiced in breeding open-pollinated corn. It was the principal breeding procedure with corn and was practiced by the farmer during harvest when selecting ears for planting the next season. In mass selection the individual plants are visually chosen for desirable traits and the grain harvested is bulked to grow the following generation without any form of progeny evaluation (Fig.3). The procedure of mass selection can be summarized as follows:

First season. From an open-pollinated source population selected ear from 50-100 plants with desired plant features are harvested and grain harvested is bulked in order to grow the following generation.

Second season. Plant harvested grain from desired plants from the previous year, selection must be practiced again with desired plant features and grain harvested is bulked.
Advantages and disadvantages of mass selection could be summarized by the following:

Advantages	Disadvantages
It can be carried out easily and simply. It is effective for improving characters with high heritability.	Pollen source is lacked of control. Ineffective for characters with low heritability.

- Gridded Mass Selection

Gridded mass selection proposed by Gardener (1961), is a procedure may be used to reduce the environmental effects (soil humidity, fertility…etc) on yield variability. In order to reduce the environmental effects, the source population is grown on small plots by dividing the land area into equally small plots or grids to ensure homogeneity of soil fertility and its humidity. Hence, this procedure gives equal representation in the mass selection from all areas of the field regardless of fertility or moisture gradients. Plants within each grid are evaluated and the superior plant in each grid selected.

- Ear-to-row

Hopkins (1899) was the first to use this method of selection based on a progeny test. This method differs from mass selection since the lines composited to form a new population are selected from progeny performance rather than phenotypic selection. The procedure of this method could be summarized as follows (Fig. 4):
First season. Select 50-100 ears from desired plant from an open-pollinated source population. Keep and shell each harvested ear separately.
Second season. With seed harvested in the previous season, plant part of seed of each selected ear in separate row in order to

evaluate each row as progeny test of selected plants. Keep remnant seed.
Third season. The population is reconstituted by bulking equal quantities of remnant seed from 10-20 lines with superior progeny performance. Plant the reconstituted population in a selected field in succeeding year and repeat the previously mentioned steps.

- Modified Ear-to-row

In modified ear-to-row selection, more accurate of progeny test is practiced by planting each selected progeny of each desirable selected plant in tow or more replications and more locations. Two years of selection are needed to complete one cycle of selection (Fig. 5). The procedure of this method could be summarized as follows:
First season. Harvest 100 ears from an open-pollinated population grown in isolated field. Keep and shelled each ear separately.
Second season. Seeds of each selected ear are planted in separate row in replicated yield trials. In an isolated field, detassel each family before anthesis. Male rows are surrounded the family rows. Males are obtained by bulking equal quantities of selected families. Depending on mean performance of each family across all locations, the best 40 progeny are selected. About the best 5 ears from each progeny are bulked to form the new population. Such a new population is used to repeat the previously mentioned procedures in the next selection cycle.

- The Recurrent-Selection Principle

Recurrent selection is any breeding system designed to increase frequency of genes for particular quantitatively inherited characters by repeated cycles of selection (Poehlman, 1987). A recurrent selection cycle involves identification of genotypes superior for the specific quantitative character being improved, and the subsequent intermating of the superior genotypes to obtain new gene combinations. Methods of recurrent selection could be summarized as:

- Phenotypic Recurrent Selection

First season. Plants are selected from the source population (Fig.6) depending on phenotype (visual selection).

Second season. Grown progenies of the selected plants and intercrossed (recombination) to obtain new gene combinations. The crossed seed is used to grow a new source population, which starts the next selection cycle. This method is most effective under low genotype × environment interaction.

- Full-sib Recurrent selection

In the 1970s CIMMYT (International Maize and Wheat Improvement Center) initiated a drought breeding program for maize using the elite lowland tropical maize population Tuxpeno Sequia (Bolanos and Edmeades 1993 a & b and Bolanos *et al.* 1993). Over eight cycles of full-sib recurrent selection for grain yield and increased flowering synchronization (reduced anthesis-silking interval [ASI] resulted in gains of up to 144 kg per hectare and per year under drought stress (Edmeades *et al.* 1999). The procedure of full-sib recurrent selection (Fig. 7) can be summarized according to Weyhrich *et al.* (1998) as follows:

Winter nursery. Within an open-pollinated population make crosses of plant-to-plant to make full-sib families.

The following season. Evaluate 100 full-sib families in replicated trails, and the best 20 families are selected.

Winter nursery. The 20 selected families are planted, and ten plants per family are selfed to produce S_1's. Self-pollinated ears from five best plants in each of these 20 selected families are bulked to represent the family for recombination.

The following season. Recombination are conducted by making $S_1 \times S_1$ plant crosses.

- Reciprocal Recurrent Selection

This method of selection is designed by corn breeders in order to improve two populations simultaneously for both general and specific combining ability. Recurrent selection for improving general combining ability must be used of a broad genetic base tester; it identifies mainly additive genetic effects. On the contrary, recurrent selection for specific combining ability depends on a narrow genetic base tester and involves both additive and non additive gene action. The procedure of this method (Fig. 8) can be summarized as:

First season. Plants are selected in each of two populations, with the selected plants of one population being selfed and outcrossed as the tester to the selected plants in the other population.

Second season. The testcross progenies are evaluated. Remnant seed from plants with superior testcross progenies are grown in the next season and intercrossed to reconstitute the two populations, completing the selection cycle.

- **Selection Environment**

Several authors (Rosielle and Hamblin 1981, Richards 1993 and Worku 2005) reported that the non-stressed selection environment is the best, because of higher heritability and expected genetic advance from selection for grain yield than the stressed environment. In contrast, numerous authors (Blum 1988, Hefny 2007, Al-Naggar *et al* 2009, 2010, 2012 and 2015, Atta 2016 and Atta *et al.* 2017) found that the target (stressed) environment is better because it would ensure the preservation of alleles for stress tolerance.

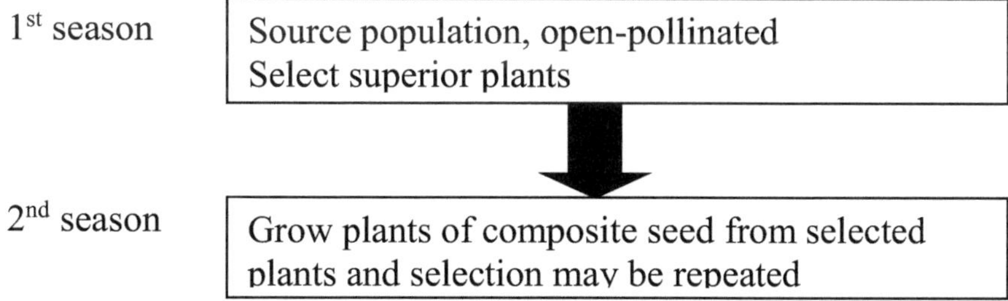

Fig. 3. Method of mass selection in corn.

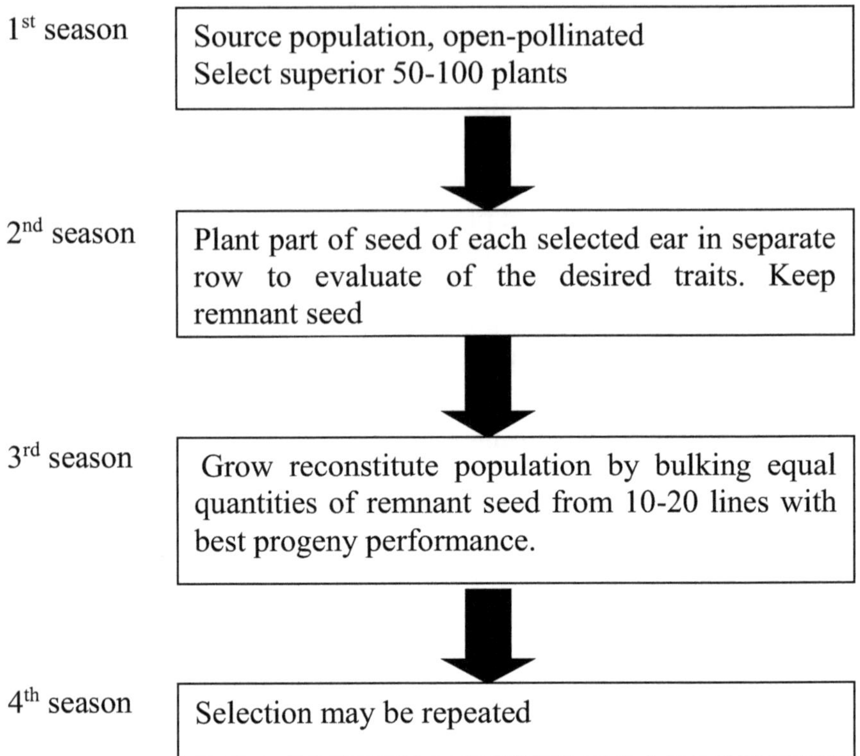

Fig. 4. Method of ear-to-row selection in corn

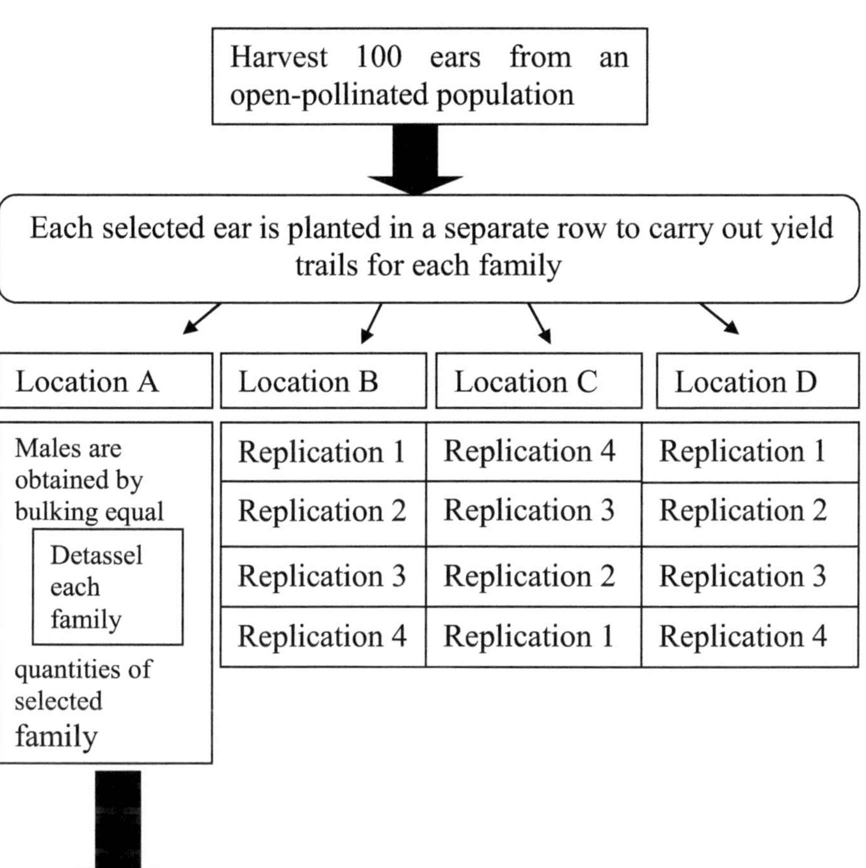

Fig. 5. Modified ear-to-row method in corn.

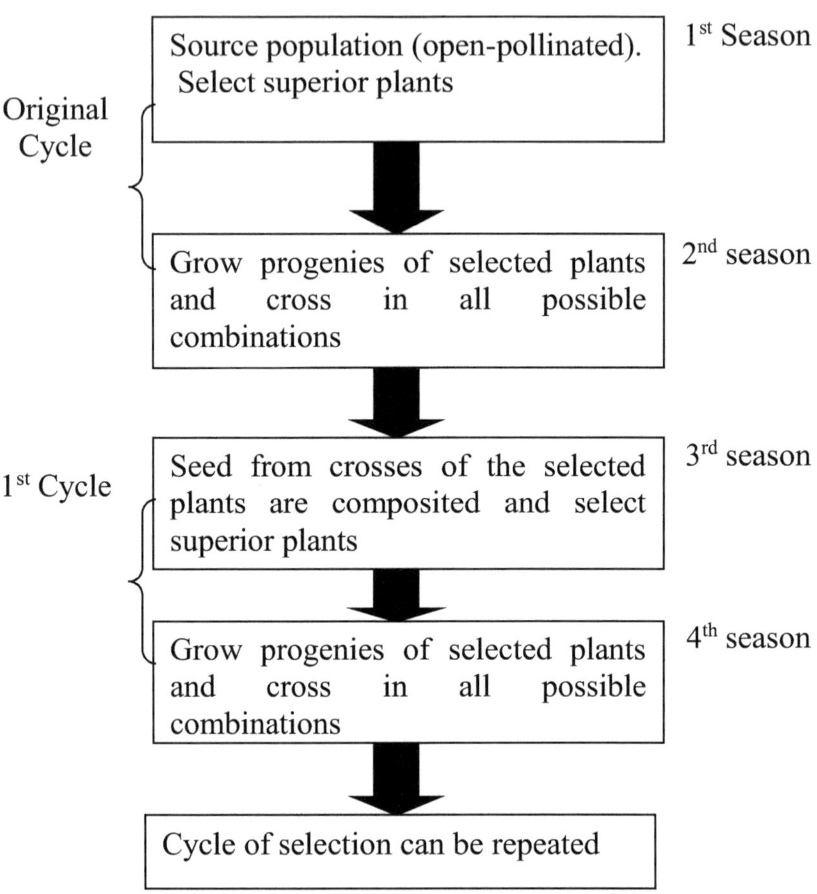

Fig. 6. Method of phenotypic recurrent selection.

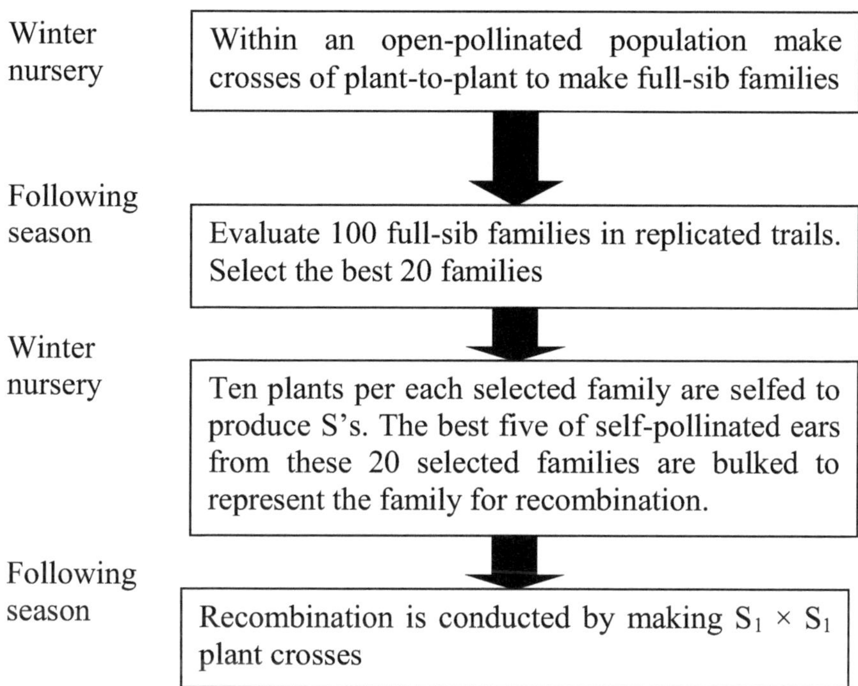

Fig. 7. Full-sib recurrent selection in corn.

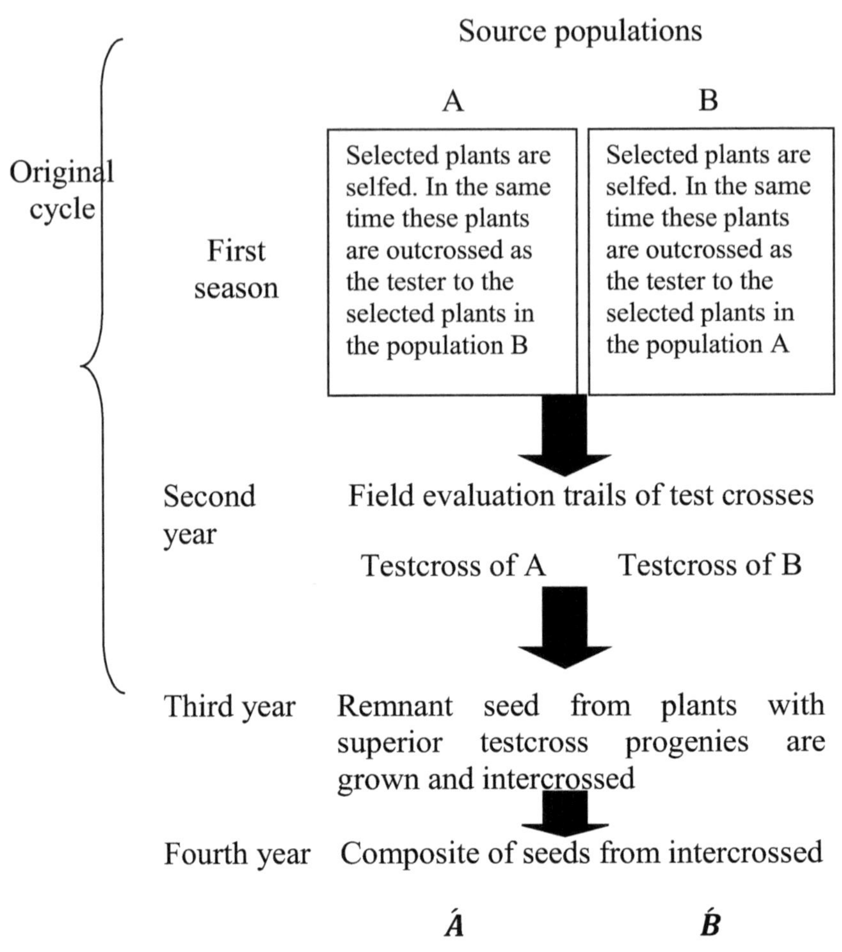

Fig. 8. Reciprocal recurrent selection.

Breeding Hybrids

Hybrid corn is the first generation (F_1) from crosses between inbred lines. The commercial use of hybrid corn depends on manifestation of abundant heterosis to justify the cost required to produce hybrid seed. Heterosis is the hybrid vigor manifested in hybrids and represents the superiority in performance of hybrids compared with their parents (Halluer *et al.* 2010). Hybrid vigor in corn is manifested in the offspring of inbred lines with high specific combining ability (SCA). Heterosis can be computed by the following equations:

$$Mid\ parent\ heterosis\ \% = [\frac{F_1 - Mp}{Mp}] \times 100$$

Where, F_1 is the first generation and *MP* is the mean performance of its parents.

$$Higher\ parent\ heterosis\ \% = [\frac{F_1 - HP}{HP}] \times 100$$

Where, *HP* is the mean performance of the higher parent.

Hybrid corn program involved four steps: (1) developing inbred lines, (2) evaluation of inbred lines, (3) yields prediction of three-way cross and double cross and (4) commercial production of hybrid seed.

- **Sources of Inbred lines**

Inbred lines may be developed from open-pollinated cultivars, composites, populations improved by recurrent selection, or populations developed by other procedures. In addition, segregating generations of single crosses, modified single or three-way crosses and double crosses.

- **Developing Inbred Lines**

1- Standard method. From a source population, a number of desired plants are selfed. In the next season the selected families are planted with ear-to-row method. Each row must be included 20-30 plants, the best elite plants are selfed. Continuous planting with ear-to-row method and continue of selfing for 5-7 generations. After 5-7 generation of selfing the inbred becomes pure lines and genetically stable.

2- Single hill method. In this method the best S_1 families are grown separately in one hill instead of one row. Each hill involves three plants. At harvest the best inbred is selected from each hill.

3- Pedigree selection. This method is commonly used in maize breeding because it proved high efficiency in genetic improvement of maize hybrids. It is similar to standard method, but the parental population used to develop inbred lines is the second generation (F_2) of elite single-cross.

4- Doubled haploids (DH). Haploid is an individuals with the gametic chromosome number (n) in its somatic cells. A doubled haploid is genotype formed when haploid cells (n) for instance egg or sperm cell undergo chromosome doubling. The resulting individuals are completely homozygous. The differences among traditional inbred and double haploids inbred can be illustrated below:

- Traditional inbred

-Produced by repeated generations of Selfing.

-In each generation, heterozygozity reduces by 50%.

-It takes approximately seven generations.

-Resulting inbred lines are highly homozygous but not 100%.

- Double haploid inbred (DH)

-DH technique is the fastest method to obtain 100% pure inbred lines.

-It takes only two generations.

- **Methods for Producing Haploids**

(1) *In Vivo*- Genetic induction. This method is widely used for corn breeding. It involves using of inducer lines. It produces high frequency of haploid generations. It is simple to operate and relatively inexpensive.
(2) *In Vitro*- Tissue culture techniques. This method used for anther culture (microspore culture). It is highly complicate and more expensive. It is greatly limited for application in breeding programs.
Production of maternal haploids using *in vivo* method requires four steps as follows:
(1) Induction of haploids by crossing the source germplasm (DH donor) as female with inducer. The inducer must be had purple aleurone and purple embryo. On the contrary, the DH donor is colorless.
(2) Identification of haploids by testing the F_1 seeds, the regular F_1 (diploid) will be purple aleurone and purple embryo. On the other hand, haploid seed will be purple aleurone and colorless embryo.
(3) Artificial chromosome doubling. Haploid plantlets are doubled by using colchicine technique. The doubled haploid is grown to maturity in the greenhouse under controlled environmental conditions.
(4) Self-pollination is required to maintain the inbred lines and for seed multiplication.

- **Evaluation of Inbred Lines**

(1) Selection through generations of inbreeding: During each generation of inbreeding, the breeder must be selected the superior lines to be selfed and continued to the next generation. Selection of the superior plants within the desirable lines is based on discarded plants that are diseased, lodged or otherwise undesirable visible characteristics. It is important that the inbred produce a good seed yield. Therefore, yield trails are conducted at early generation of the inbreeding program *e.g.* S_1 or S_2. It could be used visible plant characters that are sufficiently related to the yielding ability as a guide in selection of an inbred line contributes to its hybrid progeny. Previous studies show that only the more vigorous inbreds tend to give the more vigorous hybrid progenies. In addition, other characters such as ASI, plant height, ear size, number of barren stalks, and disease resistance should be taken into consideration during selection. During the period of inbreeding and selection, it is necessary to subject the inbred lines to drought. By such test it is possible to select the lines that will give a reliable performance under water deficit or other environmental conditions.

(2) Early testing procedure: Jenkins (1935) was the first to propose this method. This procedure can be summarized by crossing S_0 (non-inbred) or S_1 (the first generation of selfing) plants with a genetically heterogeneous tester, such as a single cross or an open-pollinated cultivar, and the performance of the testcross progeny compared in yield trials. Such procedure generally screened plants on the basis of general combining ability (GCA) to measure mainly additive genetic effects. This method enables the breeder to discard about 50% of the tested lines at early generation of selfing. At advanced generation of selfing (S_3 or S_4) the inbred lines tested on the basis of specific combining ability (SCA) by making all possible single crosses among them. SCA refers to the performance of two particular inbreds in a specific cross and measures mainly non-additive genetic effects.

- **Yields Prediction of Three-way cross and Double-cross**

The number of single-cross, three-way cross and double-cross combinations that can be made from *n* lines can be calculated from the following formulas:

$$single - crosses = \frac{n(n-1)}{2}$$

$$three - way\ cross = \frac{n(n-1)(n-2)}{2}$$

$$double - cross = \frac{n(n-1)(n-2)(n-3)}{8}$$

It can be made 45 different single-crosses, 360 three-way crosses and 630 double-crosses from 10 inbred lines. The number of crosses increases by increasing the number of inbred lines, therefore it not possible to produce and evaluate all possible three-way and double-crosses. It was necessary to find a method that can be predicted yields of three-way and double-crosses before its production. The success of yield prediction is due to the experiments' of Jenkins that published in 1934 and concluded that the predicted performance of any double-cross is the average performance of the four non-parental single-crosses that involving of the four parental inbreds. Jenkins explained that for each double-cross the gens of each line are combined with the genes of the two other lines in the corresponding hybrid. Therefore, the predicted yield for the double-cross (A×B) × (C×D) can be calculated as follows:

$$Predicted\ yield\ of (\mathrm{A} \times \mathrm{B}) \times (\mathrm{C} \times \mathrm{D}) = \frac{(\mathrm{A} \times \mathrm{C}) + (\mathrm{A} \times \mathrm{D}) + (\mathrm{B} \times \mathrm{C}) + (\mathrm{B} \times \mathrm{D})}{4}$$

In the same manner the predicted yield of the three-way cross (A × B) × C can be calculated by the following equation:

$$Predicted\ yield\ of (\mathrm{A} \times \mathrm{B}) \times \mathrm{C} = \frac{(\mathrm{A} \times \mathrm{C}) + (\mathrm{B} \times \mathrm{C})}{2}$$

- **Commercial Production of Corn Hybrids**

It can be distinguished four types of corn hybrids; single-crosses, modified single-crosses, three-way crosses and double-crosses. The production of each type can be summarized as follows:

Single-crosses: A single-cross is the F_1 progeny from a cross between two unrelated inbred lines. The inbred lines used to develop superior single-cross must be tested for their combining ability. The production technique to develop a single-cross is similar to that used in development of inbred lines; it requires hand pollination as described in artificial hybridization. The choice for inbred to be used as pollen parent, and the other to be used as seed parent, will be determined by which inbred produces plenty of pollen and which inbred possesses the highest seed yield. The single-cross hybrid is attractive for the farmers to grow, because the single-cross plants are higher in yield and uniform in appearance as well as maturity. The commercial production of single-cross seed requires three isolated fields, two isolated fields to maintain inbred line parents and isolated field to produce the single-cross seed. The inbred parents are crossed in an isolated field by planting them in separate rows. A planting pattern in common use for seed production of single-cross is one pollen parent row to four parent seed row (1:4). Another pattern in use is 1:2:1:4. In the first pattern of arrangement one-half of the seed parent rows are adjacent to a pollen parent row and in the second arrangement two-thirds of the seed parent rows are adjacent to a pollen parent row. The seed parent must be detasseled. The single-cross made from inbred lines *A* and *B* is written $A \times B$. Production of the single-cross requires three isolated fields; two for production the inbred parents and one for production the three-way cross.

Modified Single-crosses: A modified single-cross is a progeny from crossing between a single-cross as seed parent and unrelated inbred lines as pollen parent. The single-cross parent of modified single-cross is made by crossing between related inbred lines. The two related inbred lines $(A\ and\ \acute{A})$ are genetically similar to insure the minimum segregation in their hybrid progeny$(A \times \acute{A})$. The reason to use the hybrid progeny$(A \times \acute{A})$ as seed parent in the production of modified single-cross that the hybrid progeny $A \times \acute{A}$ produces higher yield than $A\ line\ or\ \acute{A}$ line alone. Two steps are required to produce the modified single-cross; the first step is production of single cross between related inbred lines $(A\ and\ \acute{A})$ by using one of them as female and the other line as male. Female parent must be detasseled. The second step is production modified single-cross by crossing the single-cross $A \times \acute{A}$ as seed parent with an unrelated line (B) as pollen parent. The modified single-cross is written$(A \times \acute{A}) \times B$.

Three-way Crosses: A three-way cross is made by crossing a single-cross and an unrelated line. The difference between the modified single-cross and the three-way cross is that the three inbred lines in the three-way cross must be unrelated. Procedures used in the production of the three-way cross are similar to that mentioned in the production of modified single-cross. The three-way cross made from A, B and C lines is written$(A \times B) \times C$. Production of the three-way cross requires five isolated fields; three isolated fields for parental inbred lines, one for production the single-cross and one for production the single-cross.

Double-crosses: A double-cross is the hybrid progeny from crossing between two single-crosses. The four inbred lines that involve in the double-cross must be unrelated. The procedures to produce a double-cross are involved crossing the inbred lines in pairs to produce two single-crosses, and then the two single-crosses are crossed to produce the double-cross. The double-cross made from A, B, C and D is written$(A \times B) \times (C \times D)$. Production of the double-cross required seven isolated fields; four fields to produce the parental inbreds, two for production the two single-crosses and one for the production the double-cross.

Biotechnology Applications in Corn Breeding Programs

In most corn breeding programs biotechnology tools continue to develop rapidly and opening new possibilities for corn breeding. Biotechnology applications according to Banzinger *et al.* (2000) will be for:

- **Fingerprinting of Inbred Lines:** The information of fingerprinting of inbred lines can be used to identify lines used as parents in a hybrid, or to predict heterosis in crosses by estimating genetic distance between parents.

- **Line Conversion:** When a trait (or traits) of interest is transferred from a donor line to a recipient elite inbred line by backcrossing; it can be reduced backcross generations from the usual 4-5 generations to around 2 by using marker-assisted backcrossing. In the same time it can be reduced the amount of linkage drag of unwanted parts that associated with transfer from the donor to the recipient line.

Marker-Assisted Selection (MAS)

Marker-assisted selection is defined as the application of molecular biotechnologies, especially molecular markers, in combination of linkage maps and genetics in order to improve plant traits. The advantage of marker-assisted selection is that opens up real prospects for new strategies in breeding because it combines conventional breeding and marker technologies. Marker-assisted selection will be an effective way to save time in breeding under drought and low-N conditions if:

- The heritability of the trait is high and field evaluation is very costly.
- Environmental effects are significant; heritability is low and if the conditions for selection are only presents occasionally (*e.g.* selection for drought tolerance in the rainy season).
- If we want to transfer a known gene to inbred lines by backcross as rapidly as possible.
Molecular markers used in marker-assisted selection allow the handling of very large numbers of genotypes during backcrossing and give the breeder the tools to reduce those numbers based on their genomic composition.

Choosing the appropriate experimental design and performing data analysis

Specific experimental designs are required for the experiments of irrigation, which the treatments would be isolated in order to avoid the interference among irrigation treatments. The present book will be discussed the experimental designs that widely used for evaluation a set of genotypes under drought stress.

- **The split-plot design**

The split-plot design and some of its application would be applied in such experiments. The split-plot design involves of main-plots, to which levels of one or more factors are applied. The main-plots are divided into sub-plots to which levels of one or more additional factors are applied. Thus each main-plot becomes a block for the sub-plot treatments; therefore the split-plot design is a special kind of incomplete block design (Steel *et al.* 1997). For example, an experiment is conducted to test factor A (irrigation treatments) with two levels *i.e.* non-stress and water stress at flowering stage in three replications of a randomized complete block design. The second factor B (genotypes) with nine genotypes can be imposed by dividing each A unit into nine sub-units and devoting such nine genotypes (B treatments) to these sub-units. It worth noting to mention that the irrigation treatments (factor A) must be isolated by making wide allays to avoid the interference among irrigation treatments.

- **Randomization and lay out of the split-plot design**

In order to deal with the previous experiment, it should be done the randomization of the studied treatments in a two-stage, first randomize levels of factor A (irrigation treatments) over the main-plots; then randomize levels of factor B (genotypes) over the sub-plots, nine per main-plot. The layout of that experiment may be as follows:

Replication 1		**Replication 2**		**Replication 3**
Non-stress		**Water stress**		**Non-stress**
8		6		5
3	Wide allay	4	Wide allay	9
2		7		2
6	4 m	9	4 m	1
9	←→	3	←→	6
1		1		8
5		8		7
4		5		3
7		2		4
		Wide allay 4 m ↕		
Water stress		**Non-stress**		**Water stress**
8		6		5
3		4		9
2		7		2
6	Wide allay 4 m	9	Wide allay 4 m	1
9		3		6
1	←→	1	←→	8
5		8		7
4		5		3
7		2		4

- **Performing Data Analysis**

Testing for normality should be done before data analysis, because the analysis of variance assumes that the studied variables are normally distributed. If they are not, then the data require transformation or some other statistically methods may be needed.

- **Testing for normality**. Testing for normality could be statistically computed by using D'Agostino test (1971) based on the *D* statistics (Table 1), which gives an upper and lower critical value. The following equation used to calculate D'Agostino test:

$D = \frac{T}{\sqrt[2]{n^3 SS}}$, where

$T = \sum(i - \frac{n+1}{2}) \times_i$

Where, *D* is the test statistics, SS is the sum of squares of the data, n is the sample size and i is the observation × rank order. Note that the *d.f* for the present test is n (sample size).

At the first the data are ordered in ascending or descending. If the calculated value falls within the critical range, then accept H_o that the sample data is not significantly different than normal distribution.

Example, data of the following example are owned to Professor Dr. M.M.M. Atta, Dept. of Agronomy, Fac. of Agric. Cairo Univ. In an experiment to evaluate nine maize genotypes under two irrigation treatments *i.e.* non-stress and water stress at flowering, by using the split-plot design in a randomized complete block arrangement with three replications. The two irrigation treatments were allotted to the main-plots and the nine genotypes were devoted to sub-plots. Data of the present example are presented in Table 2.

Table 1. Critical Values for the D'Agostino D Normality Test Taken from Zar, 1981 Table B.22

n	$\alpha =$ 0.20	0.10	0.05	0.02	0.01
10	0.2632, 0.2835	0.2573, 0.2843	0.2513, 0.2849	0.2436, 0.2855	0.2379, 0.2857
12	0.2653, 0.2841	0.2598, 0.2849	0.2544, 0.2854	0.2473, 0.2859	0.2420, 0.2862
14	0.2669, 0.2846	0.2618, 0.2853	0.2568, 0.2858	0.2503, 0.2862	0.2455, 0.2865
16	0.2681, 0.2848	0.2634, 0.2855	0.2587, 0.2860	0.2527, 0.2865	0.2482, 0.2867
18	0.2690, 0.2850	0.2646, 0.2855	0.2603, 0.2862	0.2547, 0.2866	0.2505, 0.2868
20	0.2699, 0.2852	0.2657, 0.2857	0.2617, 0.2863	0.2564, 0.2867	0.2525, 0.2869
22	0.2705, 0.2853	0.2670, 0.2859	0.2629, 0.2864	0.2579, 0.2869	0.2542, 0.2870
24	0.2711, 0.2853	0.2675, 0.2860	0.2638, 0.2865	0.2591, 0.2870	0.2557, 0.2871
26	0.2717, 0.2854	0.2682, 0.2861	0.2647, 0.2866	0.2603, 0.2870	0.2570, 0.2872
28	0.2721, 0.2854	0.2688, 0.2861	0.2655, 0.2866	0.2612, 0.2870	0.2581, 0.2873
30	0.2725, 0.2854	0.2693, 0.2861	0.2662, 0.2866	0.2622, 0.2871	0.2592, 0.2872
32	0.2729, 0.2854	0.2698, 0.2862	0.2668, 0.2867	0.2630, 0.2871	0.2600, 0.2873
34	0.2732, 0.2854	0.2703, 0.2862	0.2674, 0.2867	0.2636, 0.2871	0.2609, 0.2873
36	0.2735, 0.2854	0.2707, 0.2862	0.2679, 0.2867	0.2643, 0.2871	0.2617, 0.2873
38	0.2738, 0.2854	0.2710, 0.2862	0.2683, 0.2867	0.2649, 0.2871	0.2623, 0.2873
40	0.2740, 0.2854	0.2714, 0.2862	0.2688, 0.2867	0.2655, 0.2871	0.2630, 0.2874
42	0.2743, 0.2854	0.2717, 0.2861	0.2691, 0.2867	0.2659, 0.2871	0.2636, 0.2874
44	0.2745, 0.2854	0.2720, 0.2861	0.2695, 0.2867	0.2664, 0.2871	0.2641, 0.2874
46	0.2747, 0.2854	0.2722, 0.2861	0.2698, 0.2866	0.2668, 0.2871	0.2646, 0.2874
48	0.2749, 0.2854	0.2725, 0.2861	0.2702, 0.2866	0.2672, 0.2871	0.2651, 0.2874
50	0.2751, 0.2853	0.2727, 0.2861	0.2705, 0.2866	0.2676, 0.2871	0.2655, 0.2874
60	0.2757, 0.2852	0.2737, 0.2860	0.2717, 0.2865	0.2692, 0.2870	0.2673, 0.2873
70	0.2763, 0.2851	0.2744, 0.2859	0.2726, 0.2864	0.2708, 0.2869	0.2687, 0.2872
80	0.2768, 0.2850	0.2750, 0.2857	0.2734, 0.2863	0.2713, 0.2868	0.2698, 0.2871
90	0.2771, 0.2849	0.2755, 0.2856	0.2740, 0.2862	0.2721, 0.2866	0.2707, 0.2870
100	0.2774, 0.2849	0.2759, 0.2855	0.2745, 0.2860	0.2727, 0.2865	0.2714, 0.2869
120	0.2779, 0.2847	0.2765, 0.2853	0.2752, 0.2858	0.2737, 0.2863	0.2725, 0.2866
140	0.2782, 0.2846	0.2770, 0.2852	0.2758, 0.2856	0.2744, 0.2862	0.2734, 0.2865
160	0.2785, 0.2845	0.2774, 0.2851	0.2763, 0.2855	0.2750, 0.2860	0.2741, 0.2863
180	0.2787, 0.2844	0.2777, 0.2850	0.2767, 0.2854	0.2755, 0.2859	0.2746, 0.2862
200	0.2789, 0.2843	0.2779, 0.2848	0.2770, 0.2853	0.2759, 0.2857	0.2751, 0.2860
250	0.2793, 0.2841	0.2784, 0.2846	0.2776, 0.2850	0.2767, 0.2855	0.2760, 0.2858
300	0.2796, 0.2840	0.2788, 0.2844	0.2781, 0.2848	0.2772, 0.2853	0.2766, 0.2855
350	0.2798, 0.2839	0.2791, 0.2843	0.2784, 0.2847	0.2776, 0.2851	0.2771, 0.2853
400	0.2799, 0.2838	0.2793, 0.2842	0.2787, 0.2845	0.2780, 0.2849	0.2775, 0.2852
450	0.2801, 0.2837	0.2795, 0.2841	0.2789, 0.2844	0.2782, 0.2848	0.2778, 0.2851
500	0.2802, 0.2836	0.2796, 0.2840	0.2791, 0.2843	0.2785, 0.2847	0.2780, 0.2849
600	0.2804, 0.2835	0.2799, 0.2839	0.2794, 0.2842	0.2788, 0.2845	0.2784, 0.2847
700	0.2805, 0.2834	0.2800, 0.2838	0.2796, 0.2840	0.2791, 0.2844	0.2787, 0.2846
800	0.2806, 0.2833	0.2802, 0.2837	0.2798, 0.2839	0.2793, 0.2842	0.2790, 0.2844
900	0.2807, 0.2833	0.2803, 0.2836	0.2799, 0.2838	0.2795, 0.2841	0.2792, 0.2843
1000	0.2808, 0.2832	0.2804, 0.2835	0.2800, 0.2838	0.2796, 0.2840	0.2793, 0.2842
1250	0.2809, 0.2831	0.2806, 0.2834	0.2803, 0.2836	0.2799, 0.2839	0.2797, 0.2840
1500	0.2810, 0.2830	0.2807, 0.2833	0.2805, 0.2835	0.2801, 0.2837	0.2799, 0.2839
1750	0.2811, 0.2830	0.2808, 0.2832	0.2806, 0.2834	0.2803, 0.2836	0.2801, 0.2838
2000	0.2812, 0.2829	0.2809, 0.2831	0.2807, 0.2833	0.2804, 0.2835	0.2802, 0.2837

For each significance level, α is given a pair of critical values. If the calculated D is $\leq$ the first member of the pair, or $\geq$ the second, then, the null hypothesis of population normality is rejected.

Table 2. Grain yield per plant (g) for nine maize genotypes evaluated under two irrigation treatments

Cultivars	Replications		
	R_1	R_2	R_3
		Non-Stress	
1	212.4	223.4	203.4
2	202.4	195.8	212.4
3	224.2	205.0	200.6
4	187.8	183.8	213.8
5	180.8	209.8	183.4
6	202.8	188.8	213.3
7	212.8	186.8	201.5
8	205.0	224.8	204.0
9	216.0	198.0	194.0
		Water stress	
1	131.4	104.6	109.4
2	88.6	65.2	68.6
3	127.6	105.4	125.2
4	126.2	96.2	122.2
5	123.0	97.2	121.0
6	91.0	83.4	93.0
7	156.0	126.7	146.7
8	95.2	112.0	126.4
9	104.2	77.6	99.6

The following steps must be performed:

- Data are tested for normality by using *e.g.* D'Agostino test (1971).
- Ordered the data in ascending order.
- Calculate D statistic and compare the calculated value with critical range. If the calculated value falls within the critical range, then accept H_o which the data are not significantly different than normal distribution as shown in Table 3.

$$D = \frac{41687.8}{\sqrt[2]{54^3 \times 138820.1748}} = 0.281963$$

Comparing the calculated value with the critical range at n =54 (Table 1), n =54 is not found therefore we compare the calculated value with the critical range at n =50. Then, the calculated value falls within the critical range (0.2707, 0.2866) at 0.05 level of probability. Therefore, H_o is accepted which the data is not significantly different than normal distribution.

Then the current example data will be analyzed by using split-plot design according to Snedecor and Cochran (1994) and Steel *et al* (1997).

Table 3. Performing normality test according to D'Agostino test.

Grain yield/plant (X_i)	Rank (i)	Mean	Mean deviates squares (SS)	(n+1)/2	T
65.2	1	155.7	8198.967133	27.5	-1727.8
68.6	2	155.7	7594.799726	27.5	-1749.3
77.6	3	155.7	6107.133059	27.5	-1901.2
83.4	4	155.7	5234.25454	27.5	-1959.9
88.6	5	155.7	4508.8738	27.5	-1993.5
91.0	6	155.7	4192.322689	27.5	-1956.5
93.0	7	155.7	3937.330096	27.5	-1906.5
95.2	8	155.7	3666.078244	27.5	-1856.4
96.2	9	155.7	3545.981948	27.5	-1779.7
97.2	10	155.7	3427.885652	27.5	-1701
99.6	11	155.7	3152.61454	27.5	-1643.4
104.2	12	155.7	2657.211578	27.5	-1615.1
104.6	13	155.7	2616.133059	27.5	-1516.7
105.4	14	155.7	2534.936022	27.5	-1422.9
109.4	15	155.7	2148.150837	27.5	-1367.5
112.0	16	155.7	1913.900466	27.5	-1288
121.0	17	155.7	1207.4338	27.5	-1270.5
122.2	18	155.7	1125.478244	27.5	-1160.9
123.0	19	155.7	1072.441207	27.5	-1045.5
125.2	20	155.7	933.1893553	27.5	-939
126.2	21	155.7	873.093059	27.5	-820.3
126.4	22	155.7	861.3137997	27.5	-695.2
126.7	23	155.7	843.7949108	27.5	-570.15
127.6	24	155.7	792.3182442	27.5	-446.6
131.4	25	155.7	592.8323182	27.5	-328.5
146.7	26	155.7	81.86898491	27.5	-220.05
156.0	27	155.7	0.063429355	27.5	-78

180.8	28	155.7	627.5952812	27.5	90.4
183.4	29	155.7	764.6249108	27.5	275.1
183.8	30	155.7	786.9063923	27.5	459.5
186.8	31	155.7	964.2175034	27.5	653.8
187.8	32	155.7	1027.321207	27.5	845.1
188.8	33	155.7	1092.424911	27.5	1038.4
194.0	34	155.7	1463.20417	27.5	1261
195.8	35	155.7	1604.150837	27.5	1468.5
198.0	36	155.7	1785.218985	27.5	1683
200.6	37	155.7	2011.688615	27.5	1905.7
201.5	38	155.7	2093.231948	27.5	2115.75
202.4	39	155.7	2176.395281	27.5	2327.6
202.8	40	155.7	2213.876763	27.5	2535
203.4	41	155.7	2270.698985	27.5	2745.9
204.0	42	155.7	2328.241207	27.5	2958
205.0	43	155.7	2425.744911	27.5	3177.5
205.0	44	155.7	2425.744911	27.5	3382.5
209.8	45	155.7	2921.602689	27.5	3671.5
212.4	46	155.7	3209.432318	27.5	3929.4
212.4	47	155.7	3209.432318	27.5	4141.8
212.8	48	155.7	3254.9138	27.5	4362.4
213.3	49	155.7	3312.215652	27.5	4585.95
213.8	50	155.7	3370.017503	27.5	4810.5
216.0	51	155.7	3630.285652	27.5	5076
223.4	52	155.7	4576.773059	27.5	5473.3
224.2	53	155.7	4685.656022	27.5	5717.1
224.8	54	155.7	4768.158244	27.5	5957.2
Average **155.7**			SS **138820.1748**		*T* **41687.8**

- **Data analysis of the split plot design**

Data analysis will be done by the following steps:

*Step*1. Find the correction factor (C.F) and the total sum of squares using the data of Table 4 as follows:

$C.F = \frac{\times^2_{\dots}}{rab}$ = $(8410.4)^2/54 = 1309904$

SS total (subunits) = $\sum_{i,j,k} \times^2_{ijk} - C.F.$

SS total (subunits) = $212.4^2 + ... + 99.6^2 - C.F. = 138820.175$

*Step*2. The main-unit analysis

SS main plot = $\frac{\sum_{i,j} \times_{ij.}^2}{b} - C.F.$

SS main plot = $[(1844.2^2 + 1043.2^2 + 1816.2^2 + 868.3^2 + 1826.4^2 + 1012.1^2) \div 9] - C.F. = 123645.8393$

SS factor $A = \frac{\sum_j \times_{.j.}^2}{rb} - C.F.$

SS factor $A = [(5486.8^2 + 2923.6^2) \div 27] - C.F. = 121666.56$

SS rep. = $\frac{\sum_i \times_{i..}^2}{ab} - C.F.$

SS rep. = $\frac{2887.4^2 + 2684.5^2 + 2838.5^2}{18} - 1309904 = 1245.854$

SS error (a) = SS main plot – (SS rep. + SS factor A)
$= 123645.8393 - (1245.854 + 121666.56) = 733.434$

Table 4. Performing data analysis

Irrigations (A)	Genotypes (B)	R1	R2	R3	Total
Non-stress	1	212.4	223.4	203.4	639.2
	2	202.4	195.8	212.4	610.6
	3	224.2	205.0	200.6	629.8
	4	187.8	183.8	213.8	585.4
	5	180.8	209.8	183.4	574.0
	6	202.8	188.8	213.3	604.9
	7	212.8	186.8	201.5	601.1
	8	205.0	224.8	204.0	633.8
	9	216.0	198.0	194.0	608.0
	Total	1844.2	1816.2	1826.4	5486.8
Water Stress	1	131.4	104.6	109.4	345.4
	2	88.6	65.2	68.6	222.4
	3	127.6	105.4	125.2	358.2
	4	126.2	96.2	122.2	344.6
	5	123.0	97.2	121.0	341.2
	6	91.0	83.4	93.0	267.4
	7	156.0	126.7	146.7	429.4
	8	95.2	112.0	126.4	333.6
	9	104.2	77.6	99.6	281.4
	Total	1043.2	868.3	1012.1	2923.6

Replications	Total	Irrigations (A)	Total
R_1	1844.2 + 1043.2 = 2887.4	A1 (non-stress)	5486.8
R_2	1816.2 + 868.3 = 2684.5	A2 (water stress)	2923.6
R_3	1826.4 + 1012.1 = 2838.5	$\times_{...}$	8410.4

*Step*3. Complete the sub-plot analysis: To calculate sum of squares due to factor B (genotypes) as well as A × B interaction, together across replications it requires to establish a binary Table that combines A and B as presented in Table 5 as follows:

SS factor $B = \frac{\sum_k x^2_{..k}}{ra} - C.F.$

SS factor B = [($984.6^2 + \ldots + 889.4^2$) ÷ 6] – $C.F.$ =5378.085

Table 5. Irrigations × genotypes interaction (A × B)

Genotypes (B)	Irrigations (A)		Total
	Non-stress	Water Stress	
1	639.2	345.4	984.6
2	610.6	222.4	833.0
3	629.8	358.2	988.0
4	585.4	344.6	930.0
5	574.0	341.2	915.2
6	604.9	267.4	872.3
7	601.1	429.4	1030.5
8	633.8	333.6	967.4
9	608.0	281.4	889.4
Total	**5486.8**	**2923.6**	**8410.4**

$SS\ (A \times B) = \frac{\sum_{j,k} x^2_{.jk}}{r} - [C.F. + SS(A) + SS(B)]$

SS $(A \times B)$ =[($639.2^2 + \ldots + 281.4^2$) ÷ 3] –[1309904 + 121666.56 + 5378.085] = 5523.35

SS error $(b) = total\ SS\ (sub\ main)$ – [main plot SS + $SS(B)$ + $SS(AB)$] = 138820.175 – [123645.8393 + 5378.085 + 5523.35 = 4272.901

Degrees of freedom (*df*) for each component can be calculated as follows:

Replications, *df* = $r - 1 = 3 - 1 = 2$

Factor *A*, *df*= $a - 1 = 2 - 1 = 1$

Error (*a*), *df*= $(r - 1)(a - 1) = (3 - 1)(2 - 1) = 2$

Factor *B*, *df*= $b - 1 = 9 - 1 = 8$

$A \times B,\ df = (a-1)(b-1) = (1)(8) = 8$
Error (*b*), *df*= $a(b-1)(r-1) = (2)(8)(2) = 32$
Total, *df*= $r\ a\ b - 1 = (3)(2)(9) - 1 = 53$

Table 6. Analysis of variance of the split-plot design

S.O.V	*df*	*SS*	*MS*	*F calc.*	*F tab.*	
					0.05	0.01
Rep.	2	1245.85	622.93			
Irrigations (*A*)	1	121666.56	121666.56	331.77**	18.51	98.50
Error (*a*)	2	733.434	366.72			
Genotypes (*B*)	8	5378.09	672.26	5.035**	2.27	3.17
A × *B*	8	5523.35	690.42	5.171**		
Error (*b*)	32	4272.90	133.53			
Total	53	138820.18				

** indicate highly significant at 0.01 level of probability.

Standard errors and *LSD* of the split-plot design can be calculated for factor *A*, factor *B* and $A \times B$ interaction as follows:

Difference between factor *A*= $\sqrt{\frac{2(MSe)}{rb}} = \sqrt{\frac{2(366.72)}{27}} = 5.212$

LSD = $t_{0.05}\ (2\ df) \times Standard\ error = 4.303 \times 5.212 = 22.430$

Difference between factor *B*= $\sqrt{\frac{2(MSe)}{ra}} = \sqrt{\frac{2(133.53)}{6}} = 6.672$

LSD = $t_{0.05}\ (32\ df) \times Standard\ error = 2.042 \times 6.672 = 13.624$

Difference between two genotype means in same irrigation treatment $(A \times B) = \sqrt{\frac{2(MSe)}{r}} = \sqrt{\frac{2(133.53)}{3}} = 9.44$

LSD = $t_{0.05}\ (32\ df) \times Standard\ error = 2.042 \times 9.44 = 19.276$

Coefficient of variability: CV (a) $= \frac{\sqrt{MSe}}{\bar{x}} \times 100 = \frac{\sqrt{366.72}}{155.75} \times$

Table 7. Mean yields of maize of the current example

Genotypes (B)	Irrigations (A)		Average
	Non-stress	Water Stress	
1	213.10	115.13	164.12
2	203.53	74.13	138.83
3	209.93	119.4	164.67
4	195.13	114.87	155.00
5	191.33	113.73	152.53
6	201.63	89.13	145.38
7	200.37	143.13	171.75
8	211.27	111.20	161.24
9	202.70	93.80	148.25
Average	**203.22**	**108.28**	**155.75**
LSD(0.05)	**A=**22.430	**B=**13.624	**A×B=**19.276

$100 = 12.29\%$

CV (b) $= \frac{\sqrt{MSe}}{\bar{x}} \times 100 = \frac{\sqrt{133.53}}{155.75} \times 100 = 7.42\%$

- **The strip plot design**

The strip plot allows for greater precision in the measurement of the interaction effect. The experimental area is divided into three plots; the vertical plot, the horizontal plot and the interaction plot. The factor A is allotted to the horizontal plot, while the factor B is devoted to the vertical plot, and allow the interaction plot to settle the interaction between A and B factors.

- **Randomization and layout of the strip plot design**

In this design each block is divided into number of horizontal and vertical strips depending on the levels of the studied factors. Under the conditions of water treatments the factor A (irrigation treatments) is allotted to the horizontal plots, while the factor B (genotypes) is devoted to the vertical plots. To layout the experiment of the previously mentioned example, the experimental area is divided into r blocks (r = 3). Each block is divided into A horizontal strips (A= 2 *i.e.* non-stress and water stress). Each strip should be bordered with a wide alley (as the same of strip width) to elude interfering of the two water treatments. Then each block is divided into B vertical strips (B = 9 *i.e.* nine genotypes). B treatments are randomly assigned to these strips in each of the r blocks separately and independently. The layout of this experiment is given below:

	R1	**R2**	**R3**			**R1**	**R2**	**R3**
Non-stress (A1)	8	6	5	**Wide alley**	**Water stress (A)**	8	6	5
	3	4	9			3	4	9
	2	7	2			2	7	2
	6	9	1			6	9	1
	9	3	6			9	3	6
	1	1	8	⟷		1	1	8
	5	8	7			5	8	7
	4	5	3			4	5	3
	7	2	4			7	2	4

- **Data analysis of the strip plot design**

Data analysis of the strip plot design will be done by the following steps:

***Step*1**. Find the correction factor (C.F) and the total sum of squares as illustrated in split-plot design using the data of Table 4 as follows:

$C.F = \frac{x^2_{...}}{rab}$ = $(8410.4)^2/54 = 1309904$

SS total = $\sum_{i,j,k} x^2_{ijk} - C.F.$

SS total = $212.4^2 + ... + 99.6^2 - C.F. = 138820.175$

Step2. Compute SS due to replications and SS due to factor A (horizontal factor) by making replications × factor A table (R× A) as follows:

Table 8. Replications × Factor A (R × A Table)

	R1	R2	R3	Total
Non-stress	1844.2	1816.2	1826.4	5486.8
Water stress	1043.2	868.3	1012.1	2923.6
Total	2887.4	2684.5	2838.5	**8410.4**

SS table R × A = $\frac{\sum_{i,j} x^2_{ij.}}{b} - C.F.$

SS table R × A = $[(1844.2^2 + 1043.2^2 + 1816.2^2 + 868.3^2 + 1826.4^2 + 1012.1^2) \div 9] - C.F. = 123645.8393$

SS factor A = $\frac{\sum_j x^2_{.j.}}{rb} - C.F.$

SS factor A = $[(5486.8^2 + 2923.6^2) \div 27] - C.F. = 121666.56$

SS rep. = $\frac{\sum_i x^2_{i..}}{ab} - C.F.$

SS rep. = $\frac{\mathbf{2887.4^2 + 2684.5^2 + 2838.5^2}}{18}$ - $1309904 = 1245.854$

SS error (a) = SS table R × A – (SS rep. + SS factor A)
= $123645.8393 - (1245.854 + 121666.56) = 733.434$

Step3. Compute SS due to factor B (vertical factor) and SS due to R×B interaction *i.e.* error b by making replications × factor B table (R× B) as follows:

Table 9. Table of Replications × factor B (R× B table)

Genotypes	**R1**	**R2**	**R3**	**Total**
1	343.8	328.0	312.8	**984.6**
2	291.0	261.0	281.0	**833.0**
3	351.8	310.4	325.8	**988.0**
4	314.0	280.0	336.0	**930.0**
5	303.8	307.0	304.4	**915.2**
6	293.8	272.2	306.3	**872.3**
7	368.8	313.5	348.2	**1030.5**
8	300.2	336.8	330.4	**967.4**
9	320.2	275.6	293.6	**889.4**
Total	**2887.4**	**2684.5**	**2838.5**	**8410.4**

SS factor $B = \frac{\sum_k x^2_{..k}}{ra} - C.F.$

SS factor $B = [(984.6^2 + \ldots + 889.4^2) \div 6] - C.F. = 5378.085$

SS table R×B $= \frac{\sum_{i,k} x^2_{i.k}}{a} - C.F = [(343.8^2 + \cdots + 293.6^2) \div 2] - C.F. = 9052.91$

SS due to error b $=$ SS table R × B $-$ (SS due to factor B $+$ SS due to Reps.) $= 9052.91 - (5378.085 + 1245.854) = 2428.971$

Step4. Compute SS due to A×B interaction by making A × B table (A×B) as shown in Table 5 by the following formula:

$$SS\ (A \times B) = \frac{\sum_{j,k} x^2_{.jk}}{r} - [C.F. + SS(A) + SS(B)]$$

$$SS\ (A \times B) = [(639.2^2 + \ldots + 281.4^2) \div 3] - [1309904 + 121666.56 + 5378.085] = 5523.35$$

SS due to errror c = SS total – SS due to all other sources

$$= 138820.175 - (121666.56 + 1245.854 + 733.434 + 5378.085 + 2428.971 + 5523.35 = 1843.921$$

Degrees of freedom (*df*) for each component can be calculated as follows:

Replications, $df = r - 1 = 3 - 1 = 2$

Factor *A*, $df = a - 1 = 2 - 1 = 1$

Error (*a*), $df = (r - 1)(a - 1) = (3 - 1)(2 - 1) = 2$

Factor *B*, $df = b - 1 = 9 - 1 = 8$

Error (b), $df = (r - 1)(b - 1) = (2)(8) = 16$

$A \times B$, $df = (a - 1)(b - 1) = (1)(8) = 8$

Error (*c*), $df = (a - 1)(b - 1)(r - 1) = (1)(8)(2) = 16$

Total, $df = r\,a\,b - 1 = [(3)(2)(9)] - 1 = 53$

Table 10. Analysis of variance of the strip plot

S.O.V	*df*	*SS*	*MS*	*F calc.*	*F tab.*	
					0.05	0.01
Rep.	2	1245.85	622.93			
Irrigations (*A*)	1	121666.56	121666.56	331.77**	18.51	98.50
Error (*a*)	2	733.43	366.72			
Genotypes (*B*)	8	5378.09	672.26	4.43**	2.59	3.89
Error (*b*)	16	2428.97	151.81			
A × *B*	8	5523.35	690.42	5.99**	2.59	3.89
Error (*c*)	16	1843.92	115.25			
Total	53	138820.18				

** indicate highly significant at 0.01 level of probability.

Standard errors and *LSD* of the strip plot design can be calculated for factor *A*, factor *B* and $A \times B$ interaction as follows:

Difference between factor *A*= $\sqrt{\frac{2(MSe)}{rb}} = \sqrt{\frac{2(366.72)}{27}} = 5.212$

$LSD = t_{0.05}\ (2\ df) \times Standard\ error = 4.303 \times 5.212 = 22.430$

Difference between factor $B = \sqrt{\frac{2(MSe)}{ra}} = \sqrt{\frac{2(151.81)}{6}} = 7.114$

$LSD = t_{0.05}\ (16\ df) \times Standard\ error = 2.12 \times 7.114 = 15.10$

Difference between two genotype means in same irrigation treatment $(A \times B) = \sqrt{\frac{2(MSe)}{r}} = \sqrt{\frac{2(115.25)}{3}} = 8.77$

$LSD = t_{0.05}\ (16\ df) \times Standard\ error = 2.12 \times 8.77 = 18.592$

Table 11. Mean yields of maize of the current example

Genotypes (B)	**Irrigations (A)**		**Average**
	Non-stress	**Water tress**	
1	213.10	115.13	164.12
2	203.53	74.13	138.83
3	209.93	119.4	164.67
4	195.13	114.87	155.00
5	191.33	113.73	152.53
6	201.63	89.13	145.38
7	200.37	143.13	171.75
8	211.27	111.20	161.24
9	202.70	93.80	148.25
Average	**203.22**	**108.28**	**155.75**
LSD	**A=22.43**	**B=15.1**	**A×B=18.59**

Coefficient of variability: CV $(a) = \frac{\sqrt{MSe}}{\bar{x}} \times 100 = \frac{\sqrt{366.72}}{155.75} \times 100 = 12.29\%$

CV $(b) = \frac{\sqrt{MSe}}{\bar{x}} \times 100 = \frac{\sqrt{151.81}}{155.75} \times 100 = 7.91\%$

- **Incomplete block designs**

In many cases, a large number of corn genotypes are evaluated. In these cases, incomplete block designs are favorable for yield trails in which a large number of genotypes are being compared. An alternative set of designs for single factor experiments having a large number of treatments is the incomplete block designs. Lattice design is one of the incomplete block designs. In lattice design, the genotypes in each replication are subdivided into smaller blocks in order to reduce the experimental error due to the soil variation. Each block does not contain all treatments. With smaller blocks, the homogeneity of experimental units in the same block is easier to maintain as well as a higher degree of precision can be expected. In general, an incomplete block design is expected to give a higher degree of precision than a complete block design. Balanced and partially balanced lattice are widely used in comparing a large number of genotypes. Both of them require that the number of treatments must be a perfect square and that the block size is equal to the square root of this treatment number. The number of replications in the balanced lattice is equal the square root of the number of treatment (k) + 1 $i.e. r = K + 1$, where K is the block size. Whereas, the number of replications in partially balanced lattice is not prescribed as a number of treatments. The partially balanced lattice with two replications is referred to simple lattice; with three replications, a triple lattice; with four replications, a quadruple lattice…*etc*. (for more details see Cochran and Cox, 1957 and Gomez and Gomez, 1984).

- **Randomization and layout of lattice design**

Randomization and layout in both balanced and partially balanced lattice are the same, except for the number of replications as described above. For instance, the previously mentioned example contains of 9 treatments with three replications in each irrigation treatment. The number of treatments is square; the square root of the number of treatments equals 3, which makes 3×3 simple lattice with three replications (triple lattice). The randomization and layout is given below.

Non-stress

Blocks	R1		
B1	1	2	3
B2	4	5	6
B3	7	8	9
	R2		
B1	1	4	7
B2	2	5	8
B3	3	6	9
	R3		
B1	1	5	9
B2	7	2	6
B3	4	8	3

Wide alley

Water stress

Blocks	R1		
B1	1	2	3
B2	4	5	6
B3	7	8	9
	R2		
B1	1	4	7
B2	2	5	8
B3	3	6	9
	R3		
B1	1	5	9
B2	7	2	6
B3	4	8	3

- **Analysis of variance of the partially balanced lattice**

It is worth noting to note that analysis of variance of lattice design will be done separately for each irrigation treatment *i.e.* non-stress and water stress by the following steps:

(a) Under non-stress conditions

*Step*1. Calculate the block totals (B), the replication totals (R) and then calculate the grand total as shown in Table 12.

Table 12. Performing data analysis of partially balanced lattice

Blocks	Non-stress			Block totals (B)
	R1			
B1	212.4 (1)	202.4 (2)	224.2 (3)	639.0
B2	187.8 (4)	180.8 (5)	202.8 (6)	571.4
B3	212.8 (7)	205.0 (8)	216.0 (9)	633.8
	R2			
B1	223.4 (1)	183.8 (4)	186.8 (7)	594.0
B2	195.8 (2)	209.8 (5)	224.8 (8)	630.4
B3	205.0 (3)	188.8 (6)	198.0 (9)	591.8
	R3			
B1	203.4 (1)	183.4 (5)	194.0 (9)	580.8
B2	201.5 (7)	212.4 (2)	213.3 (6)	627.2
B3	213.8 (4)	204.0 (8)	200.6 (3)	618.4

Values in parentheses indicate to the genotype numbers

$$Grand\ total = R1 + R2 + R3$$
$$= (639 + 571.4 + 633.8) + (594.0 + 630.4 + 591.8) + (580.8 + 627.2 + 618.4) = 5486.8$$

$$Treatment\ (genotype)\ totals$$
$$T1 = 212.4 + 223.4 + 203.4 = 639.2$$

$$T2 = 202.4 + 195.8 + 212.4 = 610.6$$

$$T3 = 224.2 + 205.0 + 200.6 = 629.8$$

$$T4 = 187.8 + 183.8 + 213.8 = 585.4$$

$$T5 = 180.8 + 209.8 + 183.4 = 574.0$$

$$T6 = 202.8 + 188.8 + 213.3 = 604.9$$

$$T7 = 212.8 + 186.8 + 201.5 = 601.1$$

$$T8 = 205.0 + 224.8 + 204.0 = 633.8$$

$$T9 = 216.0 + 198.0 + 194.0 = 608.0$$

*Step*2. Compute the correction factor (C.F.), total SS, replication SS and treatment SS unadjusted as follows:

$$C.F. = \frac{G^2}{(r)(k^2)} = \frac{5486.8^2}{(3)(3^2)} = 1114999.046$$

$$Total\ SS = \sum x^2 - C.F. = (212.4^2 + \cdots + 200.6^2) - C.F.$$
$$= 4166.494$$

$$Reps.\ SS = \frac{\sum R^2}{K^2} - C.F.$$
$$= [\frac{1844.2^2 + 1816.2^2 + 1826.4^2}{9}] - C.F.$$
$$= 44.625$$

$$Treat. or\ geno. (unadj.)\ SS = \frac{\sum T^2}{r} - C.F. = .$$
$$= [\frac{639.2^2 + \cdots + 608.0^2}{3}] - C.F. = 1273.574$$

*Step*3. Calculate $Cb = M - rB$ for each block, where M is the sum of treatment totals for all treatments is appearing in that particular block and B is the block total. For example, block 3 of replication 2 contained treatments T_3, T_6 and T_9 (Table 12). Then, the M value for block 3 of replication 2 is:

$$M = T_3 + T_6 + T_9$$
$$= (224.2 + 205.0 + 200.6) + (202.8 + 188.8 + 213.3) + (216.0 + 198.0 + 194.0) = 629.8 + 604.9 + 608.0 = 1842.7$$

Then the Cb value is $= 1842.7 - (3)(591.8) = 67.3$

Values of the Cb for the nine blocks are presented in Table 13

Table 13. The C_b values for the nine blocks

Rep 1		Rep 2		Rep 3	
Block No.	C_b	Block No.	C_b	Block No.	C_b
1	-37.4	1	43.7	1	78.8
2	50.1	2	-72.8	2	-65.0
3	-58.5	3	67.3	3	-6.2
Total	-45.8	Total	38.2	Total	7.6

The sum of C_b values over all blocks (R_c) should be equal zero (-45.8 + 38.2 + 7.6 = 0)

*Step*4. Calculate the block (adj.) SS as follows:

$$Block\ (adj.)SS = \frac{\sum C_b^2}{(K)(r)(r-1)} - \frac{\sum RC_c^2}{(K^2)(r)(r-1)}$$
$$= \frac{(-37.5^2) + \cdots + (-6.2^2)}{(3)(3)(2)} - \frac{(-45.8^2) + (38.2^2) + (38.2^2)}{(9)(3)(2)} = 1574.324$$

*Step*5. Calculate the intrablock error as follows:

$$Intrablock\ error$$
$$= Total\ SS - [Rep. SS + Treat.(unadj.)SS + Block\ (adj.)SS]$$
$$= 4166.494 - [44.625 + 1273.574 + 1574.324$$
$$= 1273.971$$

*Step*6. Calculate the intrablock error mean square (*MS*) and block (adj.) *MS* as follows:

$$Intrablock\ error\ MS = \frac{Intrablock\ error\ SS}{(K-1)(rK-K-1)}$$
$$= \frac{1273.971}{(2)(9-3-1)} = 127.397$$
$$Block\ (adj.)MS = \frac{Block\ (adj.)SS}{r\ (K-1)} = \frac{1574.324}{3\ (3-1)} = 262.387$$

*Step*7. Calculate the adjustment factor μ. The equation for a triple lattice as follows:

$$\mu = \frac{\frac{1}{MSE} - \frac{2}{2MSB - MSE}}{K(\frac{2}{MSE} + \frac{2}{3MSB - MSE})}$$

Where, *MSE* and *MSB* are the intrablock error and block (adj.) mean squares, respectively.

$$\mu = \frac{\frac{1}{127.397} - \frac{2}{2(262.387) - 127.397}}{3(\frac{2}{127.397} + \frac{2}{3(262.387) - 127.397})} = 0.0562$$

Note that if *MSB* is less than *MSE*, μ is considering zero and no further adjustment is needed.

*Step*8. Calculate the adjusted treatment total T' for each treatment as follows:

$$T' = T + \mu \sum C_b$$

For example, the adjusted treatment total for treatment number 4 is calculated as follows:

$$T'_4 = 585.4 + (0.0562)(50.1 + 43.7 - 6.2) = 590.323$$

Where, treatment number 4 occupied block number 2, 1 and 3 in replication number 1, 2 and 3, respectively as shown in Table 12. Hence, C_b values as presented in Table 13 for treatment number 4 was 50.1, 43.7 and - 6.2, respectively.

*Step*9. Calculate the adjusted treatment SS as:

$$Treatment\ (adj.)SS = Treatment\ (unadj.)SS - A$$

$$A = \left[\frac{1}{MSE} - \frac{2}{(3MSB - MSE)}\right] \times [(MSE)BU - (K - 1)(MSE)(rMSB - MSE)$$

$$BU = \frac{\sum B^2}{K} - \frac{\sum R^2}{K^2} = \frac{639^2 + \cdots + 618.4^2}{3} - [\frac{1844.2^2 + 1816.2^2 + 1826.4^2}{9}] = 1661.609$$

$$A = \left[\frac{1}{127.397} - \frac{2}{(3 \times 262.387 - 127.397)}\right] \times [(127.397)(1661.609) - (3 - 1)(127.397)(3 \times 262.387 - 127.397) = 209.3694$$

$$Treatment\ (adj.)SS = 1273.574 - 209.3694 = 1064.205$$

*Step*10. Calculate the treatment (adj.) mean square (*MS*) as follows:

$$Treatment\ (adj.)MS = \frac{Treatment\ (adj.)\ SS}{K^2 - 1} = \frac{1064.205}{8} = 133.03$$

*Step*11. Calculate the F test to test the significance of treatment difference as:

$$F = \frac{Treatment(adj.)MS}{Intrablock\ error\ MS} = \frac{133.03}{127.397} = 1.044$$

*Step*12. Establish Table of the analysis of variance as follows:
- Compute the degrees of freedom for the partially balanced lattice (triple lattice) as:

$$Replication\ d.f = r - 1 = 3 - 1 = 2$$
$$Block\ (adj.)d.f = r(K-1) = 3(3-1) = 6$$
$$Treatment\ (unadj.)d.f = K^2 - 1 = 3^2 - 1 = 8$$
$$Intrablock\ error\ d.f = (K-1)(rK - K - 1)$$
$$= (3-1)[(3 \times 3) - 3 - 1] = 10$$
$$Total\ d.f = rK^2 - 1 = 3 \times 3^2 - 1 = 26$$

- Form the analysis of variance Table as:

Table 14. Analysis of variance of the triple lattice under non-stress conditions

S.O.V	d.f	SS	MS	F	
				$F_{0.05}$	$F_{0.01}$
				3.07	5.06
Replication	2	44.625			
Block (adj.)	6	1574.324	262.387		
Treatment (unadj.)	8	1273.574			
Intrablock error	10	1273.971	127.397		
Treatment (adj.)	8	1064.205	133.03	1.044	
Total	26	4166.494			

*Step*13. Estimate the relative efficiency of the partially balanced lattice relative to RCB design by calculating the effective error mean square. The average error MS may be used for comparing any pair of means, the equation of the triple lattice can be computed by the following:

$$average\ error\ MS = \frac{MSE}{(K+1)}\left[\frac{\frac{9}{MSE}}{\frac{2}{MSE}+\frac{2}{3MSB-MSE}} + (K-2)\right] =$$

$$\frac{127.397}{(3+1)}\left[\frac{\frac{9}{127.397}}{\frac{2}{127.397}+\frac{2}{3\times 262.387 - 127.397}} + (3-2)\right] = 150.669$$

Calculate the relative efficiency (*R.E*) of the triple lattice relative to RCB design as follows:

$$R.E = \left[\frac{Block\ (adj.)SS + Intrablock\ error\ SS}{r(K-1)+(K-1)(rK-K-1)}\right]\left[\frac{100}{Error\ MS}\right]$$

$$= \left[\frac{1574.324 + 1273.971}{3(3-1)+(3-1)(3\times 3-3-1)}\right]\left[\frac{100}{150.669}\right]$$

$$= 118.152\%$$

The adjusted treatment means is presented in Table 15.

Table 15. Grain yield/plant (g) for nine maize genotypes under non-stress

Treatment number	Treatment (unadj.) totals	Treatment (adj.) totals	Treatment (adj.) means
1	639.20	643.97	214.66
2	610.60	600.79	200.26
3	629.80	631.13	210.38
4	585.40	590.31	196.77
5	574.00	577.14	192.38
6	604.90	607.83	202.61
7	601.10	596.63	198.88
8	633.80	626.10	208.70
9	608.00	612.91	204.30

Grand total (adj.) = 5486.80
Grand mean (adj.) = 203.21

(b) Under water stress conditions

Analysis of variance of the triple lattice under water stress conditions can be done by repeating all previously mentioned steps as illustrated under non-stress conditions as follows:

*Step*1. Calculate the block totals (B), the replication totals (R) and then calculate the grand total as shown in Table 16.

Table 16. Performing data analysis of partially balanced lattice
Water stress

Blocks		**R1**		**Block totals (B)**
B1	131.4 (1)	88.6 (2)	127.6 (3)	347.6
B2	126.2 (4)	123.0 (5)	91.0 (6)	340.2
B3	156.0 (7)	95.2 (8)	104.2 (9)	355.4
		R2		
B1	104.6 (1)	96.2 (4)	126.7 (7)	327.5
B2	65.2 (2)	97.2 (5)	112.0 (8)	274.4
B3	105.4 (3)	83.4 (6)	77.6 (9)	266.4
		R3		
B1	109.4 (1)	121.0 (5)	99.6 (9)	330.0
B2	146.7 (7)	68.6 (2)	93.0 (6)	308.3
B3	122.2 (4)	126.4 (8)	125.2 (3)	373.8

Values in parentheses indicate to the genotype numbers

$Grand\ total = R1 + R2 + R3$
$= (347.6 + 340.2 + 355.4) + (327.5 + 274.4 + 266.4) + (330.0 + 308.3 + 373.8) = 2923.6$

$Treatment\ (genotype)\ totals$
$T1 = 131.4 + 104.6 + 109.4 = 345.4$

$T2 = 88.6 + 65.2 + 68.6 = 222.4$

$T3 = 127.6 + 105.4 + 125.2 = 358.2$

$T4 = 126.2 + 96.2 + 122.2 = 344.6$

$T5 = 123.0 + 97.2 + 121.0 = 341.2$

$T6 = 91.0 + 83.4 + 93.0 = 267.4$

$T7 = 156.0 + 126.7 + 146.7 = 429.4$

$T8 = 95.2 + 112.0 + 126.4 = 333.6$

$T9 = 104.2 + 77.6 + 99.6 = 281.4$

*Step*2. Compute the correction factor (C.F.), total SS, replication SS and treatment SS unadjusted as follows:

$$C.F. = \frac{G^2}{(r)(k^2)} = \frac{2923.668^2}{(3)(3^2)} = 316571.7$$

$$Total\ SS = \sum x^2 - C.F. = (131.4^2 + \cdots + 99.6^2) - C.F.$$
$$= 12987.12$$

$$Reps.\ SS = \frac{\sum R^2}{K^2} - C.F.$$
$$= [\frac{1043.2^2 + 868^2 + 1012.1^2}{9}] - C.F.$$
$$= 1934.654$$

$$Treat. or\ geno. (unadj.)\ SS = \frac{\sum T^2}{r} - C.F. = .$$
$$= [\frac{354.4^2 + \cdots + 281.4^2}{3}] - C.F. = 9627.861$$

*Step*3. Calculate $Cb = M - rB$ for each block, where M is the sum of treatment totals for all treatments is appearing in that particular block and B is the block total. For example, block 3 of replication 2 contained treatments T_3, T_6 and T_9 (Table 16). Then, the M value for block 3 of replication 2 is:

$$M = T_3 + T_6 + T_9$$
$$= (127.6 + 105.4 + 125.2) + (91.0 + 83.4 + 93.0) + (104.2 + 77.6 + 99.6) = 358.2 + 267.4 + 281.4 = 907.0$$

Then the Cb value is = $907.0 - (3)(266.4) = 107.8$

Values of the Cb for the nine blocks are presented in Table 13

Table 17. The C_b values for the nine blocks

Rep 1		Rep 2		Rep 3	
Block No.	C_b	Block No.	C_b	Block No.	C_b
1	-116.8	1	136.9	1	-22.0
2	- 67.4	2	74.0	2	-5.7
3	-21.8	3	107.8	3	-85.0
Total	-206.0	Total	318.7	Total	-112.7

The sum of C_b values over all blocks (R_c) should be equal zero (-206.0 + 318.7 - 112.7 = 0)

*Step*4. Calculate the block (adj.) SS as follows:

$$Block\ (adj.)SS = \frac{\sum C_b^2}{(K)(r)(r-1)} - \frac{\sum RC_c^2}{(K^2)(r)(r-1)}$$
$$= \frac{(-116.8^2) + \cdots + (-85.0^2)}{(3)(3)(2)} - \frac{(-206.0^2) + (318.7^2) + (-112.7^2)}{(9)(3)(2)} = 555.81$$

*Step*5. Calculate the intrablock error as follows:

$$Intrablock\ error = Total\ SS - [Rep.SS + Treat.(unadj.)SS + Block\ (adj.)SS]$$
$$= 12987.12 - [1934.654 + 9627.861 + 555.81$$
$$= 868.795$$

*Step*6. Calculate the intrablock error mean square (*MS*) and block (adj.) *MS* as follows:

$$Intrablock\ error\ MS = \frac{Intrablock\ error\ SS}{(K-1)(rK-K-1)}$$

$$= \frac{868.795}{(2)(9-3-1)} = 86.88$$

$$Block\ (adj.)MS = \frac{Block\ (adj.)SS}{r\ (K-1)} = \frac{555.81}{3\ (3-1)} = 92.64$$

*Step*7. Calculate the adjustment factor μ. The equation for a triple lattice as follows:

$$\mu = \frac{\frac{1}{MSE} - \frac{2}{2MSB - MSE}}{K(\frac{2}{MSE} + \frac{2}{3MSB - MSE})}$$

Where, *MSE* and *MSB* are the intrablock error and block (adj.) mean squares, respectively.

$$\mu = \frac{\frac{1}{86.88} - \frac{2}{2(92.64) - 86.88}}{3(\frac{2}{86.88} + \frac{2}{3(92.64) - 86.88})} = -0.0866$$

Note that if *MSB* is less than *MSE*, μ is considering zero and no further adjustment is needed.

*Step*8. Calculate the adjusted treatment total T' for each treatment as follows:

$$T' = T + \mu \sum C_b$$

For example, the adjusted treatment total for treatment number 4 is calculated as follows:

$$T_4' = 344.6 + (-0.0866)(-67.4 + 136.9 - 85.0) = 345.942$$

Where, treatment number 4 occupied block number 2, 1 and 3 in replication number 1, 2 and 3, respectively as shown in Table 16. Hence, C_b values as presented in Table 17 for treatment number 4 was-67.4, 136.9 and-85.0, respectively.

*Step*9. Calculate the adjusted treatment SS as:

$$Treatment\ (adj.)SS = Treatment\ (unadj.)SS - A$$

$$A = \left[\frac{1}{MSE} - \frac{2}{(3MSB - MSE)}\right] \times [(MSE)BU - (K - 1)(MSE)(rMSB - MSE)$$

$$BU = \frac{\sum B^2}{K} - \frac{\sum R^2}{K^2} = \frac{347.6^2 + \cdots + 373.8^2}{3} - [\frac{1043.2^2 + 868.0^2 + 1012.1^2}{9}] = 1515.893$$

$$A = \left[\frac{1}{86.88} - \frac{2}{(3 \times 92.64 - 86.88)}\right] \times [(86.88)(1515.893) - (3 - 1)(86.88)(3 \times 92.64 - 86.88) = 98.52$$

$$Treatment\ (adj.)SS = 9627.861 - 98.52 = 9529.341$$

*Step*10. Calculate the treatment (adj.) mean square (*MS*) as follows:

$$Treatment\ (adj.)MS = \frac{Treatment\ (adj.)\ SS}{K^2 - 1} = \frac{9529.341}{8} = 1191.1676$$

*Step*11. Calculate the F test to test the significance of treatment difference as:

$$F = \frac{Treatment(adj.)MS}{Intrablock\ error\ MS} = \frac{1191.1676}{86.88} = 13.71$$

*Step*12. Establish Table of the analysis of variance as follows:
- Compute the degrees of freedom for the partially balanced lattice (triple lattice) as:

$$Replication\ d.f = r - 1 = 3 - 1 = 2$$
$$Block\ (adj.) d.f = r(K - 1) = 3(3 - 1) = 6$$
$$Treatment\ (unadj.) d.f = K^2 - 1 = 3^2 - 1 = 8$$
$$Intrablock\ error\ d.f = (K - 1)(rK - K - 1)$$
$$= (3 - 1)[(3 \times 3) - 3 - 1] = 10$$
$$Total\ d.f = rK^2 - 1 = 3 \times 3^2 - 1 = 26$$

- Form the analysis of variance Table as:

Table 18. Analysis of variance of the triple lattice under water stress

S.O.V	d.f	SS	MS	F	
				$F_{0.05}$	$F_{0.01}$
				3.07	5.06
Replication	2	1934.654			
Block (adj.)	6	555.81	92.64		
Treatment (unadj.)	8	9627.861			
Intrablock error	10	868.795	86.88		
Treatment (adj.)	8	9529.341	1191.1676	13.71**	
Total	26	12987.12			

** indicate highly significant at 0.01 level of probability.

*Step*13. Estimate the relative efficiency of the partially balanced lattice relative to RCB design by calculating the effective error mean square. The average error MS may be used for comparing any pair of means, the equation of the triple lattice can be computed as follows:

$$average\ error\ MS = \frac{MSE}{(K+1)}\left[\frac{\frac{9}{MSE}}{\frac{2}{MSE}+\frac{2}{3MSB-MSE}} + (K-2)\right] =$$

$$\frac{86.88}{(3+1)}\left[\frac{\frac{9}{86.88}}{\frac{2}{86.88}+\frac{2}{3\times 92.64-86.88}} + (3-2)\right] = 66.44$$

Calculate the relative efficiency (*R.E*) of the triple lattice relative to RCB design as follows:

$$R.E = \left[\frac{Block\ (adj.)SS + Intrablock\ error\ SS}{r(K-1) + (K-1)(rK-K-1)}\right]\left[\frac{100}{Error\ MS}\right]$$

$$= \left[\frac{555.81 + 868.795}{3(3-1) + (3-1)(3\times 3-3-1)}\right]\left[\frac{100}{86.88}\right]$$

$$= 102.48\%$$

The adjusted treatment means is presented in Table 19.

Table 19. Grain yield/plant (g) for nine maize genotypes under water stress

Treatment number	Treatment (unadj.) totals	Treatment (adj.) totals	Treatment (adj.) means
1	345.40	345.42	115.14
2	222.40	222.83	74.28
3	358.20	359.03	119.68
4	344.60	344.74	114.91
5	341.20	341.34	113.78
6	267.40	267.09	89.03
7	429.40	428.43	142.81
8	333.60	333.89	111.30
9	281.40	280.83	93.61

Grand total (adj.) = 2923.60
Grand mean (adj.) = 108.28
After carrying out individual analysis of variance for each environment (non-stress and water stress), we have got the adjusted treatments and intrablock error. After carrying out homogeneity test by using either *F* test or Bartlett test according to Steel *et al.* (1997), the adjusted treatments and a randomized complete block model are used for combined analysis of the two

environments according to the procedure described and used by Meseka *et al.* (2006), Carena *et al.* (2009), Aharizad *et al.* (2012), Shushay *et al.* (2013) and Atta and Masri (2015) as the following steps:

*Step*1. Compute the homogeneity test by using either *F* test or Bartlett test according to Steel *et al.* (1997) as follows:

$$F\ test = \frac{larger\ MSE}{smaller\ MSE} = \frac{127.397}{86.88} = 1.47$$

Comparing the calculated *F* (1.47) with the tabulated *F* (2.98) at 10 and 10 error degrees of freedom at 0.05 level of probability, it observed that the calculated *F* was smaller than that tabulated *F*. It was concluded that the homogeneity test was not significant and the combined analysis across the two environments can be performed.

Bartlett test can be computed as follows:

Class	$d.f$	SS	S^2	lnS^2	$(n-1)lnS^2$	$\frac{1}{(n-1)}$
Non-stress (Intrablock error)	10	1273.971	127.397	4.847	48.47	0.10
Water stress (Intrablock error)	10	868.795	86.88	4.465	44.65	0.10
Totals		2142.77			93.12	0.20
pooling	20		107.139	4.67	93.40	

$$X^2 = [\sum(n_i-1)]lnS^2 - \sum(n_i-1)lnS_i^2 = 93.40 - 93.12$$

$$= 0.28\ (d.f = 1)$$

$$Correction\ Factor = 1 + \frac{1}{3(K-1)}\left[\sum\frac{1}{n_i-1} - \frac{1}{\sum(n_i-1)}\right]$$

$$= 1 + \frac{1}{3(2-1)}[0.20 - \frac{1}{20}] = 1.083$$

$$Corrected\ X^2 = \frac{0.28}{1.083} = 0.259, not\ significant$$

X^2 *with* $d.f = 1$ *is* 3.84 *at* 0.05 *level of probability.*

- **Combined analysis across the two environments**

Homogeneity test was not significant; hence combined analysis across the two environments can be computed by the following steps:

Step1. Compute the replications within environment (*E*) *SS* as the sum across *E* replication *SS*, and the pooled error *SS* as the sum of intrablock error *SS* across *E*, from the individual analyses of variance as shown in Table 14 and Table 18 as follows:

$$Reps\ withen\ environment\ SS = (Rep.SS)_{NS} + (Rep.SS)_{WS}$$
$$= 44.625 + 1934.654 = 1979.279$$

$$Pooled\ error\ SS$$
$$= (Intrablock\ error\ SS)_{NS} + (Intrablock\ error\ SS)_{WS}$$
$$= 1273.971 + 868.795 = 2142.766$$

Where, NS and WS indicate to non-stress and water stress conditions, respectively.

Step2. Establish the Table of Genotypes × Environments (G×E) in order to compute the other various *SS* as follows:

$$C.F. = \frac{G^2}{reg} = \frac{8410.41^2}{3 \times 2 \times 9} = 1309907$$

$$Environments\ (E)SS = \sum \frac{E^2}{rg} - C.F.$$
$$= \frac{5486.81^2 + 2923.60^2}{3 \times 9} - C.F. = 121667.50$$

$$Genotypes\ (G)\ SS = \sum \frac{T^2}{re} - C.F.$$
$$= \frac{989.39^2 + \cdots + 893.74^2}{3 \times 2} - C.F. = 5454.05$$

$$Table\ (G \times E)SS$$
$$= \sum \frac{X^2}{r} - C.F. = \frac{643.97^2 + \cdots + 280.83^2}{3} - C.F.$$
$$= 132452.30$$

$$G \times E\ SS = Table(G \times E)SS - (Enviro.SS + Geno.SS)$$
$$= 132452.30 - (121667.50 + 5454.05)$$
$$= 5330.714$$

Table 20. Genotypes × Environments (G×E) table

Treatment (genotype) Number	Environments		Total
	Non-stress (adjusted total)	Water stress (adjusted total)	
1	643.97	345.42	989.39
2	600.79	222.83	823.62
3	631.13	359.03	990.16
4	590.31	344.74	935.05
5	577.14	341.34	918.48
6	607.83	267.09	874.92
7	596.63	428.43	1025.06
8	626.10	333.89	959.99
9	612.91	280.83	893.74
Total	5486.81	2923.60	8410.41

Step3. Calculate the degrees of freedom for all sources of the analysis of variance table as follows:

$Environments\ (E)\ d.f. = e - 1 = 2 - 1 = 1$

$Reps/E\ d.f. = (Rep\ d.f\ )_{NS} + (Rep\ d.f\ )_{WS}\ = 2 + 2 = 4$

$Genotypes\ (G) d.f. = g - 1 = 9 - 1 = 8$

$G \times E\ d.f. = (e - 1)(g - 1) = 1 \times 8 = 8$

$Pooled\ error\ d.f.$

$$= (Intrablock\ error\ d.f.)_{NS} + (Intrablock\ error\ d.f.)_{WS} = 10 + 10 = 20$$

Step3. Establish the analysis of variance table as shown below:

Table 21. Combined analysis of variance across the two environments (non-stress and water stress

S.O.V	$d.f.$	SS	MS	F_{calc}	F_{tab}	
					0.05	0.01
Environments (E)	1	121667.5	121667.5	1135.6**	4.35	8.10
Reps/E	4	1979.279	494.82			
Genotypes (G)	8	5454.05	681.76	6.36**	2.45	3.56
$G \times E$	8	5330.714	666.34	6.22**		
Pooled error	20	2142.766	107.14			

**indicate highly significant at 0.01 level of probability.

Standard errors and *LSD* for combined analysis can be calculated for environments (E), genotypes (G) and $G \times E$ interaction as follows:

Difference between the two environments (E) = $\sqrt{\frac{2(MSE)}{rg}} = \sqrt{\frac{2(107.14)}{27}} = 2.82$

$LSD = t_{0.05}\ (20\ df) \times Standard\ error = 2.086 \times 2.82 = 5.883$

Difference between two genotypes (G) = $\sqrt{\frac{2(MSE)}{re}} = \sqrt{\frac{2(107.14)}{6}} = 5.98$

$LSD = t_{0.05}\ (20\ df) \times Standard\ error = 2.086 \times 5.98 = 12.47$

For $G \times E\ interaction$

$$G \times E = \sqrt{\frac{2(MSE)}{r}} = \sqrt{\frac{2(107.14)}{3}} = 8.45$$

$LSD = t_{0.05}\ (20\ df) \times Standard\ error = 2.086 \times 8.77 = 18.29$

Table 22. Means of grain yield/plant (g) under non-stress and water stress and the differences between the two environments

Treatment (genotype) number	Environments		Differences	Change %
	Treatment (adj.) means			
	Non-stress	Water stress		
1	214.66	115.14	99.52	46.36
2	200.26	74.28	125.98	62.91
3	210.38	119.68	90.70	43.11
4	196.77	114.91	81.86	41.60
5	192.38	113.78	78.60	40.86
6	202.61	89.03	113.58	56.10
7	198.88	142.81	56.07	28.19
8	208.70	111.30	97.40	46.70
9	204.30	93.61	110.69	54.18
Average	203.22	108.28	94.93	46.71
LSD	E=5.883	G=12.47	G×E=18.29	.....

$$Change\ \% = [\frac{(Non-stress) - (Water\ stress)}{Non-stress}] \times 100$$

- **Transformations**

The analysis of variance assumes that the studied variables are normally distributed. If they are not, then the data require transformation or some other statistically methods may be needed. The transformed data are approximately normally distributed. Three methods of transformations are commonly used according to Steel *et al.* (1997); (1) the square root, (2) logarithm and (3) inverse sine or arcsin transformation. These methods can be briefly discussed as follows:

(1) Square root transformation, $\sqrt{X}$. When data consist of small whole numbers, such as number of plants or insects of stated species in a given area, they often follow the Poisson distribution for which the mean and variance are equal. The analysis of such data is often best accomplished by first taking the square root of each observation before performing the analysis of variance. When very small values (under 10

and especially when zeros are present) are involved, $\sqrt{X+\frac{1}{2}}$ is recommended as an appropriate transformation. The square root transformation has already been recommended for percentages between 0 and 20 or 80 and 100, but the latter being subtracted from 100 before transformation.

(2) Logarithmic transformation, $\log X$. It cannot be used directly for zero values and when some of values are less than 10, it is desirable to add of 1 to each number *i.e.* $\log(X+1)$ prior to taking logarithm. In some cases, it is desired to conduct an analysis of variance where the variable is the variance. Hence, the logarithmic transformation of the variances prior to analysis is appropriate.

(3) Inverse sine transformation, $arcsin\sqrt{X}$ or $sin^{-1}\sqrt{X}$. This transformation is especially recommended when percentages cover a wide range of values, *e.g.* the percentage of barren stalks or percentage of infested plants with late wilt disease. The mechanics of the transformation require decimal fractions but tables of the arcsin transformation are usually entered with percentages. For example, the percentages of infested plant of 50% and 20% can be transformed into arcsin transformation as follows:

$$arcsin\sqrt{0.20} = Sin^{-1}\sqrt{0.20} = 26.56$$

$$arcsin\sqrt{0.50} = Sin^{-1}\sqrt{0.50} = 45.00$$

Methods of Assessment of Drought Tolerance

Assessment of drought tolerance can be carried out by using several methods as follows:

1- Stress susceptibility index (*SSI*) according to Fischer and Maurer (1978) by using the following equation:

$$SSI = \frac{1-(\frac{Y_S}{Y_N})}{1-(\frac{\bar{Y}_S}{\bar{Y}_N})}$$

2- Geometric mean productivity (*GMP*) according to Fernandez (1992) as follows:

$$GMP = \sqrt{Y_S \times Y_N}$$

3- Mean productivity (*MP*) according to Rosielle and Hamblin (1981) by the following equation:

$$MP = \frac{Y_S + Y_N}{2}$$

4- Tolerance index (*TOL*) according to Rosielle and Hamblin (1981) by the following equation:

$$TOL = Y_N - Y_S$$

5- Stress tolerance index (*STI*) according to Fernandez (1992) as follows:

$$STI = \frac{Y_S \times Y_N}{\bar{Y}_N^{\,2}}$$

6- Yield index (*YI*) according to Gavuzzi *et al.* (1997) by the following equation:

$$YI = \frac{Y_S}{\bar{Y}_S}$$

7- Yield stability index (*YSI*) according to Bouslama and Schapaugh (1984) as follows:

$$YSI = \frac{Y_S}{Y_N}$$

Where, Y_S and Y_N are stress and non-stress yield of a given cultivar, respectively. $\bar{Y}_S$ and $\bar{Y}_N$ are average yield of all cultivars under stress and non-stress conditions, respectively.
It can be applied the previously mentioned methods of drought tolerance assessment on the means of grain yield/plant by the following Table.

Table 23. Assessment of drought tolerance for grain yield/plant (g) by applying different methods

Genotypes	Environments		Change %	SSI	GMP	MP	TOL	STI	YI	YSI
	Non-stress	water stress								
1	214.66	115.14	46.36	0.99	157.21	107.33	99.52	0.60	1.06	0.57
2	200.26	74.28	62.91	1.35	121.96	100.13	125.98	0.36	0.69	0.37
3	210.38	119.68	43.11	0.92	158.68	105.19	90.70	0.61	1.11	0.59
4	196.77	114.91	41.60	0.89	150.37	98.39	81.86	0.55	1.06	0.57
5	192.38	113.78	40.86	0.87	147.95	96.19	78.60	0.53	1.05	0.56
6	202.61	89.03	56.10	1.20	134.31	101.31	113.58	0.44	0.82	0.44
7	198.88	142.81	28.19	0.60	168.53	99.44	56.07	0.69	1.32	0.70
8	208.70	111.30	46.70	1.00	152.41	104.35	97.40	0.56	1.03	0.55
9	204.30	93.61	54.18	1.16	138.29	102.15	110.69	0.46	0.86	0.46
Average	203.22	108.28	46.71	1.00	147.75	101.61	94.93	0.53	1.00	0.53

$Change\ \% = [\frac{(Non-stress)-(Water\ stress)}{Non-stress}] \times 100$. Stress susceptibility index (*SSI*), geometric mean productivity (*GMP*), mean productivity (*MP*), tolerance index (*TOL*), stress tolerance index (*STI*), yield index (*YI*) and yield stability index (*YSI*).

It was concluded that genotypes number 7, 3, 1, 4 and 5 could be regarded the most drought tolerant genotypes (Table 23). On the contrary, genotypes number 2, 6, 9 and8 could be regarded the most drought susceptible.

REFERENCES

Aharizad, S., A.K. Fard, S.A. Mohammadi and S. Sedaghat (2012). Evaluation of bread wheat recombinant inbred lines under drought conditions. Annals of Biological Research, 3(12): 5744-5747.

Almeida C., A.L. Amaral, J.F.B. Neto and M.J.C.M. Sereno (2011). Conversavação e germinação in vitro de pólen de milho (Zea mays subsp. mays). Rev. Bras. Bot. 34(4):493-497.

Almeida, C.C.S., E.P. Amorim, M.J.C.M. Sereno, Barbosa J.F. Neto, A.H. Voltz (2002). Efeito de desidratante e temperatura na estocagem de pólen de milho (Zea mays L.). In: CONGRESSO NACIONAL DE MILHO E SORGO, 24., 2002. Florianopolis. Meio ambiente e a nova agenda para o agronegocio de milho e sorgo: resumos. Sete Lagoas: ABMS/Embrapa Milho e Sorgo/EPAGRI, CD Room.

Al-Naggar, A.M.M. and M.M.M. Atta (2017). Combining ability and heritability of maize (*Zea mays* L.) morphologic traits under water stress and non-stress at flowering stage. Archives of Current Research International 7(1):1-16.

Al-Naggar, A.M.M., W.A El-Murshedy and M.M.M. Atta (2008 a). Genotypic variation in drought tolerance among fourteen Egyptian maize cultivars. Egypt. J. of Appl. Sci., 23(2B): 527-542.

Al-Naggar, A.M.M., .A.A.K., Mahmoud, M.M.M. Atta and A.M.A. Gouhar (2008 b). Intra-population improvement of maize earliness and drought tolerance. Egypt. J. Plant Breed. 12(1):213-243.

Al-Naggar, A.M.M, R. Shabana and T.H. Al-Khalil (2010). Tolerance of 28 maize hybrids and populations to low-nitrogen. Egypt. J. Plant Breed. 14(2); 103-114.

Al-Naggar, A.M.M, R. Shabana, A.A. Mahmoud, M.E.M. Abd El-Azeem and S.A.M. Shaboon (2009). Recurrent selection for drought tolerance improves maize productivity under low-N conditions. Egypt. J. Plant Breed. 13: 53-70.

Al-Naggar, A.M.M. and M.T. Shehab El-Deen (2012). Predicted and actual gain from selection for early maturing and high yielding wheat genotypes under water srtess conditios. Egypt. J. Plant Breed. 16(3); 73-92.

Al-Naggar, A.M.M., A.A. El- Ganayni, H.Y. El- Sherbeiny and M.Y. El-Sayed (2000). Direct and indirect selection under some drought stress environments in corn (*Zea mays* L.). J. Agric. Sci. Mansoura Univ. 25(1): 699-712.

Al-Naggar, A.M.M., Kh.F. Al-Azab, S.E.S. Sobieh and M.M.M. Atta (2015). Variation induction in wheat *via* gamma-rays and hybridization and gains from selection in derived heterogeneous populations for drought tolerance. Scientia Agriculturae 9(1): 1-15.

Al-Naggar, A.M.M., M.M.M. Atta, M.A. Ahmed and A.S.M. Younis (2016). Influence of deficit irrigation at silking stage and genotype on maize (*Zea mays* L.) agronomic and yield characters. Inter. J. Plant Soil Sci., 7(4): 1-16.

Al-Naggar, A.M.M., R. Shabana, S.E. Sadek and S.A.M. Shaboon (2004). S_1 recurrent selection for drought tolerance in maize. Egypt. J. Plant breed., 8: 201-225.

Al-Naggar, A.M.M., S.M. Soliman and M.N. Hashimi (2011). Tolerance to drought at flowering stage of 28 maize hybrids and populations. Egypt. J. Plant Breed., 15(1): 69-87.

Alvim, P.O. (2008). Viabilidade e conservação dos grãos de pólen de milho. 2008. 54f. Tese (Mestrado em Agronomia) – Universidade Federal de Lavras. Minas Gerais.

Athar, H.R. and M. Ashraf, (2005). Photosynthesis under drought stress. *In*: Pessarakli, M. (ed.), Handbook of Photosynthesis, pp: 793–804. Taylor and Francis, New York.

Atta, M.M.M. (2016). Genetic correlations and heritability in maize under low-N and heat stress conditions. Egypt. J. Plant Breed., 20(2): 241-260.

Atta, M.M.M. and M.I. Masri (2015). Genotypic variation among maize S_1 families of Giza-2 population under water stress conditions. American-Eurasian J. Agric. Environ. Sci. 15(5): 878-885.

Atta, M.M.M., M. Hamza and A.M. Gohar (2017). Tolerance of ten yellow corn hybrids to water deficit at flowering and grain filling. Egypt. J. Plant Breed. 21 (1):179-198.

Banzinger, M., G.O. Edmeades, D. Beck and M. Bellon (2000). Breeding for Drought and Nitrogen Stress Tolerance in Maize: From Theory to Practice. Mexico, D.F.: CIMMYT. WWW.maize_breeding_cimmyt.

Betran, J.F., D.L. Beck, M. Banziger and G.O. Edmeades (2003). Secondary traits in parental inbreds and hybrids under stress and non-stress environments in tropical maize. Field Crops Res. 83: 51-65.

Blum, A. (1988). Breeding crop varieties for stress environments. Crit. Rev. Plant Sci.;2: 199-238.

Bolanos, J. and G.O. Edmeades (1993a). Eight cycles of selection for drought tolerance in lowland tropical maize. I. Responses in grain yield, biomass and radiation. Field Crops Res. 31: 233-252.

Bolanos, J. and G.O. Edmeades (1993b). Eight cycles of selection for drought tolerance in tropical maize. II. Responses in reproductive behavior. Field Crops Res. 31: 253-268.

Bolanos, J. and G.O. Edmeades (1996). The importance of the anthesis-silking interval in breeding for drought tolerance in tropical maize. Field Crops Res. 48:65-80.

Bolanos, J. and G.O. Edmeades and L. Martinez (1993). Eight cycles of selection for drought tolerance in tropical maize. III. Response in drought-adaptive physiological and morphological traits. Field crops Res. 31: 269-286.

Borém A. and G.V. Miranda (2007). Plant Breeding. 4. ed. Viçosa: UFV. 525p.

Bouslama, M. and W.T. Schapugh (1984). Stress tolerance in soybean. Part 1: Evaluation of three screening techniques for heat and drought tolerance. Crop Sci. 24:933-937.

Campos, H., M. Cooper , J.E. Habben , G.O. Edmeades and J.R. Schussler (2004). Improving drought tolerance in maize: A view from industry. Field Crops Res. 90:19-34.

Carena, M.J., G. Bergman, N. Riveland, E. Eriksmoen and M. Halvorson (2009). Breeding maize for higher yield and quality under drought stress. Maydica, 54: 287-296.

Chapman, S.C. and G.O. Edmeades (1999). Selection improves drought tolerance in tropical maize population: II. Direct and correlated responses among secondary traits. Crop Sci. 39:1315-1324.

Chapman, S.C., G.O. Edmeades and J. Crossa (1996). Pattern analysis of grains from selection for drought tolerance in tropical maize population. In Plant Adaptation and Crop Improvement (ed. By Cooper, M. and Harmmer, G.L.).Walling Ford. UK, CAB INTERNATIONAL, 513-527.

Classen, M.M. and R.H. Shaw (1970). Water deficit effects on corn. II. Grain components. Agron. J. 62: 652-655.

Cochran, W.G. and G.M. Cox (1957). Experimental Designs. John Wiley, New York, USA.

D'Agostino, R.B. (1971). An omnibus test of normality for moderate and large size samples. BiometriKa 58(2):341-348.

Denmead, O.T. and R.H. Shaw (1960). The effect of soil moisture stress at different stages of growth on development and yield of corn. Agron. J. 52: 272-274.

Derera, J., P. Tongoona, S.V. Bindiganavile and M.D. Laing (2008). Gene action controlling grain yield and secondary traits in southern African maize hybrids under drought and non-drought environments. Euphytica 162:411–422.

Dhliwayo, T., K. Pixley, A. Menkir and M. Warburton (2009). Combining ability, genetic distances, and heterosis among elite CIMMYT and IITA tropical maize inbred lines. Crop Sci. 49:1201–1210.

Dudley, J.W. and R.H. Moll (1969). Interpretation and use of estimates of heritability and genetic variances in plant breeding. Crop Sci. 9:257-261.

Edmeadeas, G.O., M. Banzinger and T.M. Ribaut, (2000). Maize improvement for drought limited environments. In: Physiological Basis for Maize Improvement, pp: 75–111. Food Products Press, New York.

Edmeades, G.O., J. Bolanos, M. Hernandez and S. Ballo (1993). Causes for silk delay in a low land tropical maize population. Crop Sci. 33: 1029-1035.

Edmeades, G.O., J. Bolanos, S.C. Chapman, H.R. Lafitte and M. Banzinger (1999). Selection improves drought tolerance in tropical maize populations. I. Gains in biomass, grain yield and harvest index. Crop Sci. 39:1306–1315.

El-Ganayni, A.A., A.M.M. Al-Naggar, H.Y. El-Sherbeiny and M.Y. El-Sayed (2000). Genotypic differences among 18 maize populations in drought tolerance at different growth stages. J. Agric. Sci. Mansoura Univ. 25(2): 713-727.

Fernandez, G.C.J. (1992). Effective selection criteria for assessing plant stress tolerance. Proc. Intl. Symp. Adaptation of Vegetables and Other Food Crops in Temperature and Water Stress. AVRDC Publ., Tainan, Taiwan, 13-18 August, pp. 257-270.

Ferreira C.A, E.V.R Voz Pinho, P.O Alvim, V.Andrade, T.T.A.Silva and D.L. Cardoso (2007). Conservação e determinação da viabilidade de grão de pólen de milho. Rev. Bras. Milho e Sorgo 6(2):159-173.

Fischer, R.A. and R. Maurer (1978). Drought resistance in spring wheat cultivars. Part 1: Grain yield response. Aust. J. Agric. Res. 29:897-912.

Gardner, C. O. (1961). An evaluation of effects of mass selection and seed irradiation with thermal neutrons on yields of corn. Crop Sci. 1:241–45.

Gavuzzi, P., F. Rizza, M. Palumbo, R.G. Camplaine, G.L. Ricciardi and B. Borghi (1997). Evaluation of field and laboratory predictors of drought and heat tolerance in winter cereals. Can. J. Plant Sci. 77:523-531.

Gomez, K.A. and A.A. Gomez (1984). Statistical Procedures for Agricultural Research. John Wiley & Sons, Inc. New York, USA.

Grant, R.F., B.S. Jakson, J.R. Kiriny and G.F. Arkin (1989). Water deficit timing effects on yield components in maize. Agron. J. 81(1)61-65.

Hall, A.J., F. Viella, N. Chimenti and C. Chimenti (1982). The effects of water stress and genotype on the dynamics of pollen shedding and silking in maize. Field Crops Res. 5: 349-363.

Hallauer, A.R. and J.B. Miranda (1988). Quantitative genetics in maize breeding. 2nd ed. Iowa State University Press. Ames.

Hallauer, A.R., M.G. Carenz and J.B. Miranda (2010). Quantitative Genetics in Maize Breeding 3rd edition. Springer Science+Business Media, LLC, 233 Spring Street, New York, NY 10013, USA.

Hefny, M.M. (2007). Estimation of quantitative genetic parameters for nitrogen use efficiency in maize under two nitrogen rates. Int. Pl. Breed. Genet. 1:54-66.

Hopkins, C. G. (1899). Improvement in the chemical composition of the corn kernel. Bull. Ill. Agric. Exp. Stn. 55:205–40.

Jenkins, M. T. (1934). Methods of estimating the performance of double crosses in corn. J. Am. Soc. Agron. 26:199–204.

Kaefer, K.A.C., R. Chiapetti, L. Fogaca, A.L. Muller, G.B. Calixto and E.I.D. Chaves (2016). Viability of maize pollen grains in vitro collected at different times of the day. African Journal of Agricultural Research 11(12):1040-1047.

Meseka, S.K., A. Menkir, A.E.S. Ibrahim and S.O. Ajala (2006). Genetic analysis of performance of maize inbred lines selected for tolerance to drought under low nitrogen. Maydica, 51: 487-495.

Moony, H.A. and E.L. Duplesis, (1970). Convergent evolution of Mediterranean climate evergreen sclerophyll shrubs. Evolution, 24:292–303.

Nesmith, D.S. and J.T. Ritchie (1992). Short- and long- term responses of corn to a pre-anthesis soil water deficit. Agron. J. 84: 107-113.

O'Neill, P.M., J.F. Shanahan, J.S. Schepers and B. Caldwell (2004). Agronomic responses of corn hybrids from different eras to deficit and adequate levels of water and nitrogen. Agron. J. 96(6): 1660-1667.

Poehlman, J.M. (1987). Breeding Field Crops, 3rd ed. Springer Science + Business Media, New York. DOI 10.1007/978-94-015-7271-2.

Ribaut, J.M., C. Jiang, D. Gonzatez-de Leon, G.O. Edmedeas and D.A. Hoisington (1997). Identification of quantitative trait loci under drought conditions in tropical maize. II-Yield components and marker-assisted selection strategies. Theor. Appl. Genet. 94: 887-896.

Richards R.A. (1993). Should selection for yield in saline regions be made on saline or non saline soils? Euphytica 32: 431-438.

Rivero, M.R., K. Mikiko, G. Amira, S. Hitoshi, M. Ron, G. Shimon and B. Eduardo (2007). Delayed leaf senescence induces extreme drought tolerance in a flowering plant. PNAS, 104: 19631–19636.

Rosielle, A.A. and J. Hamblin (1981). Theoretical aspects of selection for yield in stress and non-stress environments. Crop Sci. 21:943–946

Salih, A.I., S. Awadalla, A. Abdelmula, M.A. Bashir, S. Baloch and W. Bashir (2014). Genetic variation of flowering trait in maize (*Zea mays* L.) under drought stress at vegetative and reproductive stages. J. Bio. Agric. Healthcare 4(20): 108-113.

Shaw, R.H. (1977). Water use and requirements of maize. A review. In: Agrometeorology of the Maize (corn) Crop. World Met. Organization Publication, 480: 119-134.

Shushay, W.A., Z.Z. Habtamu and W.G. Dagne (2013). Line x tester analysis of maize inbred lines for grain yield and yield related traits. Asian Journal of Plant Science and Research, 3(5): 12-19.

Snedecor, G.W. and W.G. Cochran (1994). Statistical Methods. 9th Ed., Iowa State Univ. Press, Ames, Iowa, USA.

Steel, R.G.D., J.H. Torri and D.A. Dickey (1997). Principles and Procedures of Statistics: A Biometrical Approach, 3rd ed., Mc Graw-Hill, New York.

Vasal, S.K., H. Cordova, D.L. Beck and G.O. Edmeades (1997). Choices among breeding procedures and strategies for developing stress tolerant maize germplasm. Proc. Symposium, March 25-29, CIMMYT, El-Batan, Mexico, Pp. 336-347.

Westgate, M.E. and D.L.T. Grant (1989). Water deficits and reproduction in maize. I. Response of the reproductive tissue to water deficits at anthesis and mid-grain fill. Plant Physiol. 91: 862-867.

Westgate, M.E. and J.S. Boyer (1985). Osmotic adjustment and the inhibition of leaf, root, stem and silk growth at low water potentials in maize. Planta 164: 540-549.

Weyhrich, R.A., K.R. Lamky and A.R. Hallauer (1998). Responses to seven methods of recurrent selection in BS11 maize population. Crop Sci. 38:308-321.

Worku, M. (2005). Genetic and crop-physiology basis of nitrogen efficiency in tropical maize. Ph.D. Thesis. Fac. Agric. Hannover, Univ. Germany.

Xiong, L., R.G. Wang, G. Mao and J.M. Koczan (2006). Identification of drought tolerance determinants by genetic analysis of root response to drought stress and abscisic acid. Plant Physiol.;142:1065–1074.

Contents

Title	Page

yes
I want morebooks!

Buy your books fast and straightforward online - at one of the world's fastest growing online book stores! Environmentally sound due to Print-on-Demand technologies.

Buy your books online at

www.get-morebooks.com

Kaufen Sie Ihre Bücher schnell und unkompliziert online – auf einer der am schnellsten wachsenden Buchhandelsplattformen weltweit!
Dank Print-On-Demand umwelt- und ressourcenschonend produziert.

Bücher schneller online kaufen

www.morebooks.de

SIA OmniScriptum Publishing
Brivibas gatve 1 97
LV-103 9 Riga, Latvia
Telefax: +371 68620455

info@omniscriptum.com
www.omniscriptum.com

Printed by Books on Demand GmbH, Norderstedt / Germany